Mettlework

Nettlework

A MINING DAUGHTER ON MAKING HOME

JESSICA E. JOHNSON

CINCINNATI 2024

Acre Books is made possible by the support of the Robert and Adele Schiff Foundation and the Department of English at the University of Cincinnati.

Printed in the United States of America

ISBN-13 (pbk) 978-1-946724-75-5
ISBN-13 (e-book) 978-1-946724-76-2

Designed by Barbara Neely Bourgoyne
Cover photo used with permission of the author.

The press is based at the University of Cincinnati, Department of English and Comparative Literature, McMicken Hall, Room 248, PO Box 210069, Cincinnati, OH, 45221-0069.

Acre Books books may be purchased at a discount for educational use.
For information please email business@acre-books.com.

CONTENTS

For Ingrid and Anders

And that feeling of "we're right here" that must be held,
the way you carry a brimming pot so nothing gets spilled.

—Tomas Tranströmer, "Baltics" (Patty Crane translation)

• • •

Because I always feel like running.
Not away, because there's no such place.

—Gil Scott-Heron, "Running"

A NOTE TO THE READER

This book is a work of memory, though parts of it take place before I was alive or before my experiential memories took shape. Stories of my family's time in Alaska, Leadville, Slick Rock, and Granite filtered down to me through my mother, forming a kind of inherited memory, seen through the lens of another person.

We make and remake the memories through which we understand ourselves, and in the process of iteration, the memories change. After my daughter was born, I learned that the version of the past I'd learned from my mother—her account of birthing and raising children alongside my father's mining career—didn't always match the letters she wrote at the time. The gap between story and document started me down the path of writing this book.

Even in creating a work of memory rather than a documentary or work of scholarship, I set myself the task of investigation. To write places and events that took place before my own memory, I relied on my mother's correspondence from the times and places in question as well as present-day interviews with her. I also read what I could about the places where my family's story intersects with published facts and accounts. I often found that the details of the family version fit with what I learned from research, but at times I also had to read the family story against the grain of recorded history. I tried to stay true to the facts where they were available and aimed to not reproduce anything I knew to be false.

In service of memory and metaphor, and out of respect for people who didn't consent to being part of my retelling, I wrote with varying degrees of specificity. I changed or abbreviated some names and made some figures into composites. "The Company," for example, was in fact several companies, in keeping with the monolithic way I remember them. There's an implied point here, too, about the nature of capitalism as experienced in the frame of this story, the way all the companies from a child's vantage point might as well have been the same.

Speaking of frames, this book takes place in the so-called American West, a place shaped by violence and trailing a well-worn popular mythology that serves to render the violence invisible or, if visible, natural. Because it's my story, this book takes place during a time and in a culture in which that mythology was so pervasive that it was itself difficult to see. And as the book is a work of memory rather than a documentary or work of scholarship, the story's frame is sometimes necessarily very close. My intention in writing a close-framed story set in this time and place is not to prop up the Western images and ideas whose harms I've come to understand, but to illuminate their construction from within the intimate configurations of family and the imagination of a single life. I offer this story modestly and socially, as one story among many, and with the knowledge that countless so-far-untold stories of this place are needed.

ANTECEDENTS

Silver and gold have a price. At this writing, $24 and $1,936 per troy ounce, respectively. Silver conducts electricity and heat better than any element on earth. Films glazed or studded with silver salts and exposed to light turn into pictures as the silver splits from chloride or bromide, becomes pure Ag, and darkens. Silver has antimicrobial properties; on nineteenth-century wagon trains, settlers prevented illness by drinking from milk bottles that held silver coins. Highly reflective, silver powers solar panels and slicks the backs of mirrors. A *silver bullet* is a solution so effective it might not be real.

And gold keeps its luster. It's unsullied by the air. Gold jewelry and bricks fill Dubai vending machines. Gold salts injected in the synovial space soothe inflamed joints. Gold leafs buildings and covers chocolate and floats at the bottoms of schnapps bottles and patches troubled teeth. In language, it replaces virtue and safety: *you're golden, good as gold, golden boy, golden years.*

Metals come from mountains, where the still-hardening rock once held metal-rich brines that heated and cooled, rose and receded, traveled through cracks, reacting all the way—and left behind silver, gold, copper, lead, and their alloys.

Veins are one structure this process makes. They stretch into the hard formations, filling fissures. Their shape is often less like a tube and more like a hand, whole families of them like jagged sheets of paper laid through the mountains. This is the non-commodity shape of metal: embedded in its surroundings and existing for its own sake as part of the earth's story.

The plane that marks the difference between metal and nonmetal is called the *selvedge*, a word for the divide created by the nature of a thing itself: the self edge.

As mountains weather, metal flakes away, runs with the rain, and collects in the slowest parts of a river. This is often the first way people take the metal out of the earth: they squat to sift the stream for what's visible. And when no more is visible, they dredge the river bottom. And when they've removed the river gold, they begin to tunnel. They crush rock to extract, through first physical and then chemical means, substances with unusual shine.

I was born into my parents' restlessness and freedom-seeking, arriving at a time when they were always on the move from mine to mine, encampment to encampment. In 1975 they started their married life together in Alaska's backcountry, on a site they had to be flown into. In a place without roads, they were the only two people for many miles. My future father had a degree in mining engineering, but it was not an engineering job. They were making nominal improvements to a string of claims so the owner could keep his lease and deduct the small price of their labor from his taxes.

Both my parents had grown up watching *The Lone Ranger*, *Sergeant Preston*, *Sky King*, and *Bonanza* in regular houses on regular streets in small western cities. They were young and healthy, with only themselves to worry about. The remoteness, what they thought of as *adventure*—the possibility of a wild and free *good life* amid and in search of elements—was the point.

In the first version of the world I understood, we didn't hold on to more than we could fit in a pickup. There were no other kids, seldom any women besides my mother. There were no parks, no libraries, no swimming pools, no play dates, no grocery store, no doctor, no babysitter. There was no telephone, no television, no movie theater, no gym, no extended family, no sidewalk. There were the same books, over and over. There was inside; there was outside. There were miners. There was us.

In the years to come, as metals prices rose and fell and mines went bust or became untenable or started up elsewhere, my father took on all kinds of mine work. He sought a degree of autonomy and wound up with increasing responsibility. He worked for bigger companies and smaller; he worked for himself, and then for companies again.

My mother's job was to follow. She set up camp, sourced parts when things broke down, ran errands, fetched water, and fed who needed feeding. When I was born, and then my brother, it was hard for her to leave camp, and so we endured together in improvised places not built for anyone's thriving.

To grow up beside mines the way I did is to grow up inside a story, without other stories to compete. It's a story that has room for men who get things done, for women who make do and ask for little, for children who ask for little and do not dream, for children content to grow into the narrow space their parents inhabit. It's a story about what has value. It's a story about what and who has consequence. It's a story made for profit, not for care. A story in which something better is always a little farther down the tunnel, over the next hill, or embedded in the next property.

We had one way of understanding why we were wherever we were: because a mine was there. Eventually one camp was bad enough—buggy, cow-trampled, my brother ill with giardiasis—that we moved to "town" (Baker City, Oregon) while my father continued to work away from us. We didn't stay long before moving again to North Idaho's silver district, a place with a relatively stable, productive mine. Then finally, when my father became an executive, we settled in a rural suburb near the company's corporate office, though my father then began long years of mine-to-mine travel on his own. Our trajectory traced a jagged path from what's now a designated wilderness to an isolated cul-de-sac.

Even as our lifestyle edged closer to typical, remnants of a more provisional existence decorated our suburban house: a cast-iron pan salvaged from Alaska; enamel camp dishes as everyday dinnerware; two chairs made, like snowshoes, of bent wood and hide; my mother's proud refusal of kitchen gear that required electricity; the way she reveled in occasional power outages; my parents' preference for traveling light and accumulating little; their regard for any place we were as circumstantial, inherently temporary.

In *Don't Let Me Be Lonely*, Claudia Rankine writes, "We carry the idea of us around with us." The camp years underlay the way my parents understood themselves and the way they organized our lives.

When I was grown, living in a rapidly gentrifying city and struggling to establish the place-bound and predictable kind of life my parents had fled, I'd sometimes run into family friends who mentioned the tent years, the mine-following. *Oh my goodness, the last time I saw you, you were about this high and living in a tent! In Granite!* I'd smile and nod. This, I believed, was part of my parents' story but not my own.

Wasn't I making a different kind of home, building a different kind of partnership? Didn't things that had *already happened* simply crumble away into a bluish realm, untouchable? I imagined I was making *progress*—linear movement through a field. I imagined I'd gotten *out* and *away*, and that was all that mattered.

But the camp years came back to find me during another period of narrowing. When my daughter was born, they were waiting. They had a shape: a buried tangle of fear and absence arising, paradoxically, from the tasks of care and the truth of love.

I found myself walking the floors of my small city house, in camp but not in camp, this time as mother instead of child.

THE POLAROID BABY & THE SHAPE OF TIME

2020

Clementine and Paul at the long table. Their lovely wide foreheads, large eyes set wide, small ears tipped just a little outward. Soft spring light on the window sides of their faces.

Paul picks a strawberry from the plate, chews quietly.

"*Mom*. Do we have any sugar?" he asks.

"Probably."

"Can I have some. On this?"

"It's fruit, honey. It's already sweet."

I don't want them to be always imagining the sweeter. I want them to love what the earth grew just as it is. I want the ready sweetness I washed, sliced, and placed in front of them to be enough.

They want and they need and they ask, more than one body can answer.

They give, more than one body can hold.

Their infinite intelligence. Their infinite seeking. Their infinite curiosity. Their *Did you know, Mom? Did you know? That a coconut is a fruit and not a nut at all?*

Their joy is bigger than any joy I can remember. Their sadness is the size of the whole house. Their occasional anger—at a limit set or come against—can spin the room. They long for infinite comfort. They want the sunrise-colored past before consciousness.

Mom they intone when they have sensed my concentration turning away from them, even if their father, kind and attentive and warm, is right there and available.

Mom: upon arriving home, before spilling out the day's events.

Mom: the first word that comes to them at the end of a summer night when they've been up late and wild and they want it to go on forever.

Mom: the person they seek after bedtime when the quiet floods their minds with hyper-specific fears and regrets.

Mom: the word and the body between them and what they don't yet understand as oblivion.

And what becomes of the grown human being who has to hold the vastness of children? The body stays present, but the person can't always withstand the constant pressure blurring the edge of self.

In the dailiness of the children, the suddenness of them, the endlessness of them, I peel off a consciousness and live in her.

Not always: sometimes, with Paul's teary cheek on my dry cheek, his arms locked tight around my neck, his legs gripping my hip, all of me can be in unison with *the child, the child, the child,* his warmth and smell.

Sometimes I can simply listen. Sometimes I can say *I'm sorry* for the ways I may have let my tiredness show.

Who is standing here right now filling the sink with soap and hot water, wrapping a cut onion in an old plastic bag and nestling it in a fridge drawer? My body has faded into the house tasks, chopping what will be this meal and then the next, wiping down the table and counter that will be dirty in an hour, putting away each object that will come out again, maintaining this handed-down domestic world that's supposed to make *home*.

I'm living inside an unspoken elsewhere. I'm thinking about work, about people I know, about later and otherwise. I'm thinking about the story I've been trying to tell about our family. I'm forming words and not saying them.

Paul is talking; he is looking up from a book he's too young for: "*Mom*! I *said* what is a Kafkaesque dream?"

I orient myself, decide how much to say.

"Mom."

"Just a weird one. That's all it means."

Silence. An unsatisfying answer.

"But why is it *called* that?"

Clementine has set her strawberry down, shifted her eyes toward the window, one leg curled up, the other swinging. She is eight and already showing the here-not-here quality that people called—at least as far back as my great-grandmother—*reserve.* This word means holding much of one's self back.

Clementine, born first, a girl, a beloved self-outside-myself, arrived with open eyes. I struggle to find words that can hold us. To refashion the story I was given into something that can keep her whole, so she won't have to hold so much back. To learn and relearn what it means to taste the sweetness that is *here.*

For Clementine, and for the child I was before I grew into this blurred self. The child who learned early how to rise into her head when things got hard, a child uneasy in space and time.

• • •

As I was born into my parents' restless search for metals, Clementine was born into our struggle to cohere and contribute in a place where Kevin and I thought we could stay.

St. Johns was a neighborhood on a peninsula extending from central Portland toward the convergence of two rivers. Portland: puffed up on its parks and trees, its green plans and policies. Swollen on its quirkiness, its local and precious. Even before quirkiness, certain of its virtue.

We'd come to Portland seven years before, in 2004, familiar with its brand identity but mostly ignorant of its history. We'd hoped to find work and a place we could afford. We'd hoped to find something to belong to.

We'd come because Kevin was from there—or close enough, the southwest suburbs—and because I loved my mom's hometown, or at least the idea of it I'd formed from family photos and short visits and her descriptions of a secure and comfortable life she'd left without realizing she'd never come back for good. When she and I were living on a mine site in mud and crushed rock, Portland was *the city*, the only one imaginable. It was bruised petals and wet sidewalks. It was the place where my mother and I seemed less out of place.

We'd come because Kevin had a team of doctors in Portland who were familiar with his complex diagnoses and how to manage them. For a young man, he had a combination of autoimmune illnesses so rare that his case is documented in medical literature. After a few scary episodes, his doctors had arrived at a drug cocktail that was working. During the two years we'd been together, he'd been mostly well—enough that I didn't really know what it meant for him to be ill.

We'd come from Seattle, a city less boastful and more openly ruthless, better educated and more expensive—a place that didn't do so much to sell a vision of itself as the center of an easygoing DIY *good life*. Like many young and educated others, we'd bounced over the big river in our U-Haul on a bright day when the freeway roses were blooming, cautiously imagining that a warmer, drier, smaller place might embrace us.

We hadn't known that Portland had long been in the business of attracting people but not so much in the business of sustaining them. We looked forward to cheaper rent and more parking. We didn't anticipate our many counterparts arriving from elsewhere. We didn't anticipate how few places there would be for us to land.

Having held the city so long in my mind as a refuge, I couldn't see it as what it was.

And when we arrived, my thoughts about any place started with what I could gain from it.

Portland did not embrace us. But even with our meandering careers, our too-small and gig-ish jobs, we saved up over a period of years and, with help from family in the last days of easy credit, signed the papers on a small house. I remember a local writer of my parents' generation saying, of our purchase, "You stupid, stupid people," his own no doubt lovely home having been purchased in less expensive times.

The house had tall ceilings and windows and a wide front porch two strides from the street. The uninsulated walls and original glass formed a porous barrier to the neighborhood. Inside, we could hear the zoom of an errant Pontiac or Mercury or Kia and voices from the sidewalk: a man talking on his phone, stirring up his own agitation, raucous kids laughing and taunting. The derelict ones returning from the Lucky Mart with a telltale weight in a black plastic bag, greeting the neighbors' dog by name. The untrimmed Airedale mix flopped around his front yard and barked at every passerby until his human stuck her head out and told him to stop. She'd know the name of the person walking and shout sorry and hello.

On warm Friday evenings we'd drive off the edge of the peninsula, across its iconic green bridge, and out of town, toward a beach on the big river. In May and June, cottonwood tufts floated like snow in the sky. We'd throw the ball for the dog, who would plunge out into the river and return to the spot, return to the spot, sleek as a black seal. Big barges came down the river, massive in proximity, their containers meant for the ocean. Kiteboarders' sails sighed against pinkening skies.

Many things hadn't turned out like we thought they would on the day we crossed the bridge, but I loved that house. A Swedish immigrant had built it between 1908 and 1910. He moved the house—had it transported—a few years later from a location about a mile from our lot after winning a settlement from the railroad company that had destroyed his access to a major road. He picked up his little hand-built home and set it down again. After him, through the middle of the twentieth century, a German immigrant lived in the house's four small rooms with his wife and two sons. I imagined the boys sharing the hot, street-facing room that was my office before it held Clementine's crib. One son, a shipbuilder, lived in the house a long time and appeared to have fixed things badly, fixed things with what he could find, fixed some things not at all. The house had a story—of people holding on to a place, making things work inside some old and not entirely good bones, arranging as much beauty as they could within its walls. I loved that it was just enough space and not more. I loved its small aesthetic considerations. I loved it because it was a place meant for staying.

Before Clementine in the St. Johns house, the main gravitational field was between Kevin's body and my body. Mine was healthy, spinning out energy and sometimes too much worry and ill feeling about what I thought were my shortcomings and mistakes. His was one day ill and feverish, ill and exhausted, ill and sweaty, or the next day well and moving heavy shovelfuls of dirt around the tiny yard. He was often singing, holding guitars in his arms when he was too weak to do anything else. My body was always about some business, chopping something, sweeping something, transplanting something. His broad body loved my body unfailingly. He folded around me when he could stand up and also when he couldn't.

My alarm went off at 4:00 a.m. He'd rise later and wander in with coffee.

"Whatcha doin'?" he'd ask, leaning down and kissing my head, and I'd say, "Writing." He needed no more detailed answer than this as he shifted away to his guitar or typewriter.

Before Clementine, I wrote nearly every morning, like it was exercise. I wrote as well as I could the shape of the cat sitting on the back of the couch, the spiderweb outside the living room window. If it was going well, I felt the connection between word and word, sentence and sentence, like the connection between breath and breath, the objects around me kindling. If I could keep it *there* without veering into anxieties and to-do lists, this kind of writing was meant to redeem the rest of the day's frustrations and boredoms, worries and frictions, the too many too-small jobs.

It did, and it also didn't. In my workmanlike determination to rise early and devote myself to something, as my father had devoted himself to the mine, I described and described, because in doing so I did not have to show up on the page, did not have to implicate myself, did not have to answer or relate except as an observer. I hated writing *I* and found ways not to.

I'd take afternoon breaks from the home office to walk past the midcentury dry cleaner, past the tall Catholic church clad in pink stucco whose doors were often open, whose constellated candles I could sometimes glimpse in the blue interior. I wrote as if I could exist in objects. I had no language for who and where I was. I let the written world contain my blue longing to take some kind of shape.

Finally, six years after we first crossed the big river as would-be Portlanders, we had reason to feel like things were starting to pan out. I had stumbled into community college teaching, a kind of work I was surprised to like. For the first time, it was a job I didn't want to leave or even clock out of. It was full of students who were funny, interesting, and smart. They didn't know what not to say and didn't always do the work, but they also didn't expect things to be easy. It was full of the city that you don't see in ad campaigns, the one made up of people rather than commodified claims to virtue. Kevin finally had a second master's degree in library science and full-time hours at an art college library small enough that he had touched—and possibly read—every book in the collection.

In the house, we'd replaced the peeling kitchen linoleum and torn through the bathroom's layers of tile and wallpaper to the plaster and horsehair beneath, then down to the wooden box the Swedish immigrant had no doubt nailed together with his own hands one hundred years before. We had put back wallboard and wiring and tile and paint and fixtures, all now in order and more or less to code. The dog was no longer a puppy but a trained fetching athlete, solid muscle and controlled energy. We were starting to feel viable.

Is it any wonder that one late summer, we thought it might be safe to test the limits of our expansion and make way for a new person, someone at once *of us* and wild, unknown?

In September, the month of plate-size dahlias and dry golden lawns, during a break from creating documentation for my fall classes, I peed on the absorbent white part of a home pregnancy test and watched the second blue line appear. I walked to a nearby school field to throw the ball for the dog.

I came home and lay on the bed. I looked up at the fan on the ceiling. Kevin was at work. I called my mom.

"Exciting," she said with a quaver. "I remember that feeling. Are you scared?"

"Totally. I mean. I have no idea how it will work out." I was verbally gesturing at all sorts of math, no longer theoretical: hours worked and slept, dollars earned and spent, economies of creativity and responsiveness.

"Everybody feels like that," she said. "It does work out."

My mother was not the only one who would smooth specific concerns into vague generalities. As my body grew and my capacity for action dwindled, my questions became sharper than anyone's answers. I saw my doctor to confirm the test. I hadn't been her patient for long. A woman about my mother's age with feathered silver hair, she asked me if my husband had a job. I said yes and wondered what that had to do with anything. One job of the kind our degrees could get us, here, would not be nearly enough.

One afternoon, I found myself at a darkening trailhead an hour outside of town. Four months in, I was just past the fabled transition to the second trimester. According to books and mom lore, I was not supposed to feel sick and tired by this point in the pregnancy.

To drive along the big river east of Portland is to travel through the artifacts of cataclysm and erosion and public administration: a massive hydropower operation that flooded the oldest continuously inhabited village in North America, concrete walls that make the salmon's journey home infinitely harder, newer white windmills keeping lazy time along the skyline. On the Oregon side, trailhead signs line the old highway which runs adjacent to the freeway. The trails follow, through rock, fern, and waterfall, the creeks that feed the big river.

Walking a trail, you might cut along a basalt cliff face under a fringe of dripping ferns; you might reach out and rest your hand on a round rock formation, rust and silver, without knowing it's the fossilized stump of a twenty-million-year-old tree, entombed by mudflows from an old eruption. I'd hiked there often in prepregnant times, knowing nothing of ancient conifers.

On this day I knew only the need to crawl out of my body, which became a need to get out of the city, to see some salmon even from inside a massive salmon obstacle, to focus on something outside myself.

I was trying to get to the Bonneville Dam, with its fish ladder and viewing window, but I missed the exit, having misplaced it in my mind as I now misplaced many things, inconsequential and otherwise. As I pulled off the freeway to turn around, I knew I didn't have it in me to stop at the dam. It was too late; late in the season and late in the afternoon. I'd wasted too much time getting out of town. I was suddenly too tired to imagine myself walking very far at all. I wasn't going make it.

I pulled into a trailhead parking lot to rest. Just for a minute, I told myself. I swung the door open and found the ground. Hiking was out of the question, but I imagined I'd sit by the creek and try to wake up. At least my unfamiliar body might forget itself for a few minutes, I thought, before it was time to drive back into town.

I ate rarely, vomited frequently, and hungered for basic instructions. Where was the manual—the real one, the good one, the one that would tell me what to do with so many too-small jobs I couldn't quit or call in sick to, with writing and writing, even without any purpose I could understand. What was I supposed to do with this endless need to live in language and the deep well of my own attention?

No book warned me that in the early days of pregnancy I would sometimes be so tired that I couldn't make it through a forty-minute commute, that I would have to pull into a parking lot and lapse into involuntary sleep. No brochure explained that I would feel so nauseated that no part of my waking existence would be pleasant. The internet showed me that the baby was the size of a walnut, then an apple, and so on. No one said, *You won't know who you are.*

One of the famous books I was given featured a cast of pencil-drawn, bloused women from 1993. They had downcast Mona Lisa smiles and neat bellies. I wanted someone who had walked this path before to tell me not just *that* it would be okay, but *how*.

As I stepped along the edge of the pavement, my eyes combed the broad shallows where the light was leaving the creek and the air, turning both the same black. It would be too dark, too late.

And then a flash: a light patch. A powerful thing splashing the dark.

Enormous fish held steady positions in the creek bed, undulating hard, their backs visible above the surface. No longer silver, but black and red and covered with patches of white death where their bodies had already begun to break down. Ragged ocean soldiers. They moved on each other and tussled, competing for the place in which to lay their genes. They were so close, so exposed that I imagined reaching down and grabbing one.

They keep going, I thought. *Their bodies start to decay while they're alive. They keep going. They are a vehicle for the next generation and nothing more and it's okay. They change and change and change and then they die. That's how.*

It's not that I believed I would die having the baby, though beneath the cheery aesthetics of maternity land, women do, even still, even here.

It's that the baby, even unborn, brought a kind of darkness to the surface of life, where I could see only what was immediately in front of me. In this darkness I was alone, though others had clearly been there first, having left, with their lives, this very narrow passage. In this darkness a being inside me was growing up into my lungs, already troubling my breathing. In this darkness, I was losing the ability to do many of the things, nearly all of the things, through which I could feel myself. In this darkness, I was becoming, in my own mind, unrecognizable.

In the cold, gray months before Clementine was born, I'd rest my unfamiliar heaviness on the cold back steps and call my mother. I wanted a distraction, but I couldn't keep my worry from rising into the conversation.

My questions were mundane and concrete as I tried to plan how I'd parcel out hours and minutes and dollars and cents. "Diapers," I said. "What are you supposed to do about diapers?" All the options seemed expensive or time consuming or both, and my mind was gnawing on the environmental impact of each choice, a factor that would have been decisive for my rapidly fading old self.

"Well, when you guys were little," she said, "I didn't have running water, right, so cloth wasn't an option. Disposable diapers were pretty new, and we couldn't really afford them, so we had these cheap ones that leaked and we were always running out because we couldn't get to town, so you just didn't wear them a lot of the time."

Just not wearing them a lot of the time, I knew, was not going to work. I'd stopped running and lifting weights months before and was having trouble finding the motivation even to walk every day.

"It does make me feel better. And I know I should. But every time I do it, it feels like one more thing that's going to be impossible after I have the baby. Normally with working out, you get a little better at it all the time. But now it gets harder and harder, then pretty soon . . ." Pretty soon, I would be able to do very little. A belief had surfaced from somewhere deep that pregnancy was the beginning of a long and narrow path on which there's no space for even basic needs. I couldn't see the end of it well enough to imagine a time when my fitness would return.

"I remember when I was pregnant with your brother," my mom said, "and we were living in this trailer on the side of a mountain, and in the spring, everything was just mud. If I stepped in it, I'd sink way down in. The doctor in town kept scolding me about not getting enough exercise, and I told him I was trying, but I just thought *you don't understand*." She paused. "I guess what I'm saying is, things won't always be perfect. For you or the baby."

She had only her own life to offer. Everything would work out, she was saying, if I could be endlessly fluid, endlessly gentle, addressing my needs by finding a way to need less. Things would be fine if—for the first time—I could find a way to be more like her. A familiar feeling grew from these conversations: the soft shame of being alone in the desire for something different.

One of the medications that kept Kevin upright and walking and fixing things around the house and moving dirt around the garden, a strong anti-inflammatory he'd been on for a decade, spontaneously stopped working, and he started to stumble, started needing to lie down. He saw all his normal specialists and new ones. I held his hand as he gritted his teeth through a bone marrow biopsy. As the doctor drove a nightmarish needle into Kevin's lower back, the physician assistant who was supposed to be helping couldn't stand to watch and walked out of the room.

The marrow they biopsied turned out to be unremarkable; the hematologist could not explain why Kevin had too few white blood cells to qualify for any of the new, highly effective treatments for his illnesses.

I searched *environmental impact cloth diapers.*

I searched *washing cloth diapers.*

I searched *side effects* plus the names of each semi-poisonous medication Kevin was offered.

I searched *parenting with* and the names of Kevin's diseases.

I searched *heritable* and the names of Kevin's diseases.

I searched *budget cloth diapers.*

I searched *health insurance babies.*

I created one giant spreadsheet.

As far as Kevin's health, it turned out there was nothing for this new state of affairs, no alternative treatment, no story we could use to understand it. It just was. We would have to bend again, to reconfigure ourselves in response to something hard.

He went to his parents' house to get one of his dead grandpa's canes. A nurse told us that we'd better start submitting disability paperwork soon, before things got worse. Kevin said, "It's not time for that yet," and kept going to work and half-smiled when the muscles in his face were weak.

I taught as much as I could in an all-out effort to be among the adjuncts who worked enough to qualify for health insurance, positioning myself near a garbage can in case I had to throw up.

In my night class, the students threw me a surprise baby shower. They brought me packages of diapers, fuzzy fleece blankets, and tiny pink dresses. They smiled politely at the name we chose, and some shared the customs of childbirth and recovery from their home countries.

One woman, holding her can of Monster, leveled her eyes with mine and said, "Don't mess around. Get the epidural." An eighteen-year-old father urged me to watch *The Happiest Baby on the Block* by Harvey Karp and demonstrated his swaddling technique with a jacket and a textbook. They made me feel like it might work out by showing me exactly how.

Just before Clementine was due, Mom came to stay. Each night, she and Kevin and I sat around our small dining table, eating a meal she'd made if I was too tired, the anticipation growing as the days and then weeks stretched past the due date. Mom and I went to the movies. Mom and I went to the Japanese garden. I was done driving; I couldn't fit behind the steering wheel. Mom drove me to appointments where measurements confirmed that the baby was still okay.

And one afternoon, she waited in the car while I hauled the waterlogged vessel of myself to a job interview for a full-time teaching position on my favorite campus among the several I taught at. I puffed up a flight of stairs and answered a few easy questions as the interviewers smiled. I told them that I wanted to be part of the life of the college. That I had something to contribute, not just to students, but to the conversation about the problem of their perpetual failure, which was really *our* failure. I had things to say about how we could collectively do better by them. My answers were fine, but it was already an old self answering the questions, a hard seed buried inside a swollen body, the swollen body lodged in a span of days between past and the future.

And then she was here.

In the summer after Clementine was born, I couldn't easily fit myself into the texture of house and street, the pieces of broken story and conversation that I heard passing on the sidewalk. I spent new days inside: bouncing, nursing, bouncing, and singing along with the Talking Heads. I hit repeat on "This Must Be the Place (Naive Melody)," a song about wanting to be home and already being there.

Minutes into hours of watching the baby and waiting for her next move. Hours into days of tending, days into weeks without direction.

Kevin came home at the end of the day, and Clementine brightened to his voice. He held her as tenderly as I did, changed her diapers—laundered them—and loved her in feelings and actions. And still our experiences were splitting. We were two people who loved each other, partners, and then we woke up as Mommy and Daddy in a Mommy-Daddy-Baby house where he went away all day—clean and tidy and tired—and my existence was inner and immediate and sticky and unpaid. Attached to the baby, I became a person laden with work, and, I feared, no longer fit to earn.

* * *

This mother-life. To whom could I tell it?

To whom could I tell these shapeless days in shapeless pants?

To whom could I tell the hours in the hospital, Clementine's eerie intelligence, my immediate perception of her personhood, her strangely open eyes?

To whom could I tell the morning in the recovery room when I got the email that the job was mine just as I was moving with difficulty from the small bed to the one chair, trying not to bleed or leak on the relatives, mostly Kevin's, who came in a steady stream to meet our baby?

To whom could I tell the morning in the recovery room when, wondering at the seeming infinity of Clementine's existence, I calculated the year she might graduate from high school, the range of years in which she might have children herself, and then, perversely, the year in which, if she were incredibly lucky and healthy, her life might end. To whom could I tell the terror of understanding what it meant that—in the best case—I would precede her in death? That there would be thirty-odd years of her life that I would never see before she too passed back into the oblivion from which she came. And that these mornings were the same morning.

To whom could I tell the warm afternoon when we waited anxiously for discharge papers, nervously buckled Clementine into her seat, and drove home from the hospital on the highway called Sunset, down the hill called Sylvan, with forest rising on either side of us, the city's white and brick towers rising above a canopy of green—heading toward the city she'd be from?

In a blue notebook that I was terrified I would lose, I wrote only to the person who had lived inside me, the one whose existence felt like a reckoning. I wanted to preserve the substance of us in words.

I wrote her skin and hair and the shape of her eyebrows. I wrote her ladylike hand gestures, the way she spread her fingers, matching palms together or putting the back of one against her cheek.

I wrote of driving with her through the long tunnel to Beaverton, the white tile and the rows of lights and remembering, suddenly, making the same drive with my own mother when I was little and Mom was pregnant with Andy, and Grandma lived on SW 90th. I wrote about the game we played: holding my breath through the tunnel, then the sharp exhalation when we shot out into the sky, the end of the tunnel's echo, the whole experience like diving down and reemerging.

I wrote to her a little every day. I wrote that we didn't know each other yet, though I'd been carrying her for months. I wrote, *I want to tell you everything I know.*

She changed and changed. I struggled to keep track, as if in writing I could make our existence something solid. None of my tiny letters closed.

And soon enough, anxiety crept in: my worry about what was still undone at the end of every day, my worries about returning to work. Underneath all of it was the fear I couldn't bear to speak: that it might not be possible to be both a mother and a distinct, coherent person with a little bit of agency and awareness.

Reading the journal now, I notice the way I turned my attention from the baby to the shape of the tree outside the window, the cat on the back of the couch, going back to the familiar habit of taking refuge in observation. And then I was simply chronicling my fatigue, my fear about going back to work. I tried to tell us it would be all right, as if I was writing to myself instead of her. Under general and baseless self-reassurance, the here-now of the particular baby disappeared.

One afternoon I was lying on our tiny armless loveseat with Clementine sleeping faceup on my chest. She had been in the world for six weeks. She grasped my index finger in her whole hand, riding the waves of my breathing, dream expressions flickering across her face. I could see her brows draw together and drift apart. I was not asleep, was not doing anything but smelling her thick patch of hair. I watched the ceiling fan chop, felt the small box of our hundred-year-old living room vibrate, the periodic thwack of the dog's tail on hardwood.

It was still new, this feeling of working constantly and accomplishing nothing. Still new: waiting for her next need, never quite able to settle. I did not feel like myself. I wiggled a hand under her head and slid the other along her tender spine, eased my body out from under hers. I nestled her into the plastic mechanical baby swing whose presence mocked our curated vintage grownup living room.

I checked email. One from my mother: a couple of scanned Polaroids and two scanned letters.

Look familiar? she wrote. Meaning the similarities between Clementine's face and mine, the sameness of features, the openness of expression. My baby and the Polaroid baby linked by a genetic off-rhyme. Different nose, same mouth.

The Polaroid baby wore handknitted booties and rested on a crocheted blanket against a backdrop of the wood-paneled seventies. My mother's profile and long, elegant fingers hovered over.

The letters sprang open into squares of light named by year and location. *SlickRock1979.pdf. Granite1980pregnant.pdf.* I recognized my mother's light cursive. In places a toddler's scribble—my own decisive points and vortexes—overwrote her lines.

In the weeks after Clementine's birth, family-made blankets, some books that my mom had managed to save, these letters—spaces and objects and stories I thought I'd left behind—had been gathering around my daughter as if they'd been waiting for us all this time, up around a bend. I was wrong about the shape of time: I'd left nothing behind. I'd been walking in a circle I could only now perceive.

In the first letter, I was eighteen months old and my parents were living in a tent in the desert outside Slick Rock, Colorado. They had been drilling for silver, but the deposit Dad prospected in his first and last venture with his own two-person company turned out to be little more than dust.

This particular time in the desert is a high point in my mother's retelling. She describes it as a scenic idyll where the stars came out after the Coleman lamp's slow shutoff hiss. In the desert, she has said, a still-cold watermelon brought from town to a place with no refrigeration, tasted sweeter. It was just the three of us then, and I fit relatively easily into their strange life.

She opens the letter in present tense: the two of us waking up together, washing in cold water, eating the last of a good cantaloupe, *opening the tent flaps to let in the day's last cool breeze.* Writing to her own faraway mother, she describes us as adventurers amusing ourselves in a beautiful nowhere.

Dad's the one who can move earth and shift the situation. She writes about how he and his partner are taking on drilling contracts to pay off the debt they incurred to start up the company, hoping their ancient equipment makes it through. He's thinking about what to do next, considering almost any option: taking a job, saving up to try again, figuring a way out of mining.

He works around the clock. Mom's and my plan for the day, besides the usual chores of wood chopping, meal making, and dish doing, is going to the river for a swim. She was without a job, without deadlines, without material conveniences, and with men's needs to consider as primary. But she had a certain kind of freedom: time and time and time.

At the end of her long passage about their prospects—*his* prospects—she writes, *We're awfully tired of being unsettled.* Even then—in the best times. I knew how much a refrain this would become, how much unsettlement followed.

The feeling of being *right here,* a sense of presence as if we were in the only place that mattered, faded in and out like a radio signal.

After the first few weeks with Clementine, I could see, could feel, could almost taste and smell how extreme my mother's situation was. A lump in my throat.

I knew how often she might not have had what she needed, for herself or for my brother and me. Our isolation, just beginning here, was no longer part of the past, no longer theoretical.

I opened the second letter. It was dated about nine months later, March of 1980. I'd just turned two, and my mom was heavily pregnant with my brother. All winter we'd been living between a tent and a small, iced-in trailer outside an alpine ghost town in eastern Oregon. My father was managing a small mine owned by a New York financier, trying to rebuild savings after the loss in the desert. The trailer had water, but the pipes unfroze for only an hour a day in the late afternoon, and only if she ran the heat full blast. If we needed to make the ninety-minute trip into town, for groceries or medicine or a laundromat, Dad had to clear snow off several miles of road.

This letter, too, opens with a scene, this time a desperate one: *It's snowing, and snowing,* my mother writes, *and snowing all week long and Jessie and I are confined to the indoors.* She describes me trying to persuade our dog to go on an imaginary expedition to the North Pole. I give up and flop down on top of the dog's aging body—a motion I remember—resting with my only friend in the intractable boredom of days and days in the same room.

Like in the Slick Rock letter, Mom uses present tense. But this time she's writing herself out of the immediate situation. I recognize the way that writing can be both presence and avoidance.

In flashback mode, she tells of her return from a visit to Portland, her hometown full of blooming flowers. She's allowed herself to imagine grass and window boxes for the new cabin, along with water that flowed freely from the tap. It would have been my parents' first house. When she pulls up to the expanse of gravel she cleared in anticipation, *Garbage was strewn all over as well as scrap lumber, plastic, all sorts of propane bottles, broken tools. After I worked so hard to clean it up before I left I burst into tears on the spot. It looked like a combined junkyard and used car lot. Inside didn't cheer me much either.* Whoever was doing the work left the mess in a place without garbage service or a nearby dump. The plumbing isn't in after all. I start wailing about going back to Portland. She burns my grilled cheese sandwich and the heat from the stove melts the plastic flowers the Company put in the kitchen. The Company has left the light bulbs bare but saw fit to pay for a *microwave oven,* she notes, which she can't imagine as useful if it can't bake bread.

By the end, her despondency submerges under a fog of optimism. Vague resolve clouds the particularity of her writing. *I feel pretty good about everything. We may even get water soon.*

Here, I could see her: what she went through and what she told herself about it. The gaps between expectation and reality, needs and events, her comfortable upbringing and her adult surroundings. The tendency to patch these spaces over with resolve, even when they were chasm-deep. I could see her positioning and repositioning herself, naturalizing or minimizing the hardships of a life not meant for homemaking. She was the person who endured without complaint by altering her expectations. The self she rebuilt was one who could be okay with the way things were.

In the email to which the letters were attached, my mom said, "There are stacks and stacks. This is just a couple."

Clementine blinked awake, interrupting whatever attention I'd been able to pay to anything. *Look familiar?*

I moved to get her, inside my own making-do, the everyday no-place of me and the beloved, unsettling baby.

When Mom called or visited—when anyone called or visited—I cleaned up and smiled and said I was *doing fine.* It was not necessarily a lie or an evasion. But there was no one to ask what *fine* should look like now.

I no longer sought my mother's prescriptions. I no longer said most of what I was thinking—to her or anyone.

We gazed together at the baby. I sat, never not waiting for the baby to need something.

Sometimes I could taste a little piece of time. At night, when we threw open the windows and doors to cool the house—after tidying the living room of blankets and burp rags, after making dinner, after putting Clementine down—I'd say *see ya* to Kevin and take my laptop to the darkening backyard: the spice of our rock roses, a little exhaust, hints of cottonwood from the river. Here, if I was lucky, I could make a scene of this domesticity, even if I had nowhere to place myself but inside of it.

Untethered from the nursing infant for an hour or two, I sat on the porch listening to how thin the walls were in the neighborhood, how close together we were in our shelters, how slim the boundaries between our lives.

I was not following a man from mine to mine, trying to make temporary homes in impossible places. But I could feel the downward curve of my shoulders. I was growing into the shape my mother made when her posture bent toward the needs of others.

My body had taken to caring for Clementine easily. Every act of tenderness overlay fear that I'd be swallowed by the imperative of loving her, that I would cease, that the entity I was used to knowing as *myself* could not continue under the force of this love.

If I could speak to the young woman my mother was when she was writing letters out of camp, what would I say?

I can feel your life closing around me. Where is the space for Clementine and me both? How do the two of us grow at once? I don't know the terms to help me say it. I can't find the frames to help me see it. Right now, away from her for the first time all day, I'm imagining a dog bounding over a field. I want to run out into the night.

• • •

This is what I remember of our encampments.

A tent frame made of two-by-fours and pale canvas. Studying the squares made by shadows and sunlight on its walls. A summer night in a trailer, Dad home from work covered in dust and suddenly pressed into parental duty, falling asleep in the middle of a book. I remember a clear spring sky against fir trees—fiercely blue—and wondering if it was deep, if it had a boundary, if it was a substance. Baby Andy pounding on a window in the morning as a bulldozer headed to the mine site. The smallness of my own boot measured against a bulldozer's tracks through heavy mud.

The mine smelled like damp stone and men smelled like diesel when they stood around outside our tent: talking shop, disregarding us when I wanted nothing more than to be central, part of some activity that mattered after a long day playing alone.

A whiff of chlorine from a motel swimming pool smelled like a built environment, people, possibility. From our hours-long rattling over gravel into town for groceries and drinking water, I remember the cursive allure of Coca-Cola signs, the high, orange rotating ball at the 76 station and how I imagined myself curled up inside it. Girls bearing towers of soft-serve ice cream smiled at us as they leaned out the windows of the drive-in. Their hair feathered away from their faces and their eyelids sparkled crystal blue and purple.

Back in camp, I dreamed myself into their existence, using the sight of them, Beatrix Potter and Winnie the Pooh stories, and a Juice Newton song from Baker City AM radio as raw material for a fantasy social life involving tea parties, eyeshadow, and rodeo queens. As early as I can remember, I woke up longing for other places.

I remember the entwinement of Dad and work, of men and work. He left and left and left for work, and his work was the deep cause of everything we experienced. I remember wishing to grow up to be him instead of her.

I imagined myself into a future where I wouldn't be the one left in the making-do. Ours was an extreme setting, a narrow sort of social ecotone, and it didn't take much for me to become unnatural, impossible, strange. A girl who wanted to be something else, a kind of woman she'd never seen.

In my memories of camp, my mother is almost never an image so much as a feeling of immensity that I traveled near and needed to stay close to. I longed for changes in the weather—for snow, for sun, or for the ominous darkness of a coming thunderstorm. I remember how her changes could surprise me, how much a kind of providence she was—coming up with a song to break my boredom, her voice turning harsh in what seemed like a sudden loss of patience.

Looking at the shape of the pages, the lines on the paper, the tilt of my mother's light handwriting, I could remember what it felt like to watch her. I could remember one particular sky-blue and grass-green day just after we'd moved to town.

She sat on a low lawn chair, balancing a writing pad on her thighs. I orbited. I was shirtless, wearing jeans that had belonged first to a boy, and though I had learned by then that I was a girl, I was okay with it only when I could imagine that *girl* meant omnidirectional wild potential, something like flame. The particular paper, the particular pen, were not routinely available to me. They belonged to her only, as almost nothing did. They lived in the drawer below the telephone, conductor of messages, bringer of news, conduit of exciting voices.

It was new to have a telephone and to live in a house, near stores and neighbors, and not in a camp near men and a mine. It was new, and I was afraid it wouldn't last.

In this new life I wanted to proclaim, to make words, to say and say.

My father was somewhere working and had been somewhere working forever. He was working at the mine whose location and distance I wasn't certain of. "The mine" was in fact not one but many. I only knew that it was the most important thing, the reason for leaving places I wanted to stay or staying places I wanted to leave.

If he was coming home on Friday night, from the lilac-scented dawn of Friday, events would coalesce around his arrival: cleaning up, baking bread, getting the boring things done before he was here.

His arrival would be an event. My mom, me, the baby, the dog would listen for the rush of his big approaching engine, gather on the porch, and watch the new-used, root-beer-colored pickup pull into the shade of locust trees that I found painfully tall and graceful, like green celebrities in a warm sky. He would push open the heavy door and step down from the cab happy.

After he left again, all the waves made by his presence slowly diminished back into the large body of undifferentiated time.

On this day, I remember, my mother was writing a letter, and that was the solid thing, an activity generated by my mother's own wish to do it. It was something she did by herself, for her own reasons, less often now since we moved down to town from the tents and trailers at the mine site. When she wrote, she seemed hard and away, looking for words. Maybe imagining. Maybe depicting. Maybe confiding things that couldn't be told to me.

"When can *I* write?" I would ask, stepping in and out of the house's shadow. I wanted my mother to turn from her own thoughts and take my dictation, to make important the words I wished to make important.

"This is my letter, honey. Please let me finish. You can get your art supplies from inside."

But I was usually tired of drawing, tired of all I couldn't give shape to, tired of the motions of my body in its search for some kind of business.

• • •

The business that mattered was the mine, and the mine was always out there, mysterious and mostly unimaginable. Mines were the deep *why* of our existence, but they were not the domain of a wife or child. We had a limited understanding of how they worked even when Dad was working them. We almost never went near them, almost never went inside.

Just a handful of times, I've been in mines, ones Dad worked that I can barely recall and antique ones on elementary school field trips that I remember better. They were damp, and the rock smelled cold as we lurched along a small railroad inside a mountain. When we stopped, a headlamp shone on a wet wall where a vein stood out from the undifferentiated gray surrounding it.

So much tunneling and infrastructure: for a slightly shining strand. Seeing how the vein ran through the earth, the logic of mining seemed simple and severe. In terms of volume, it required the removal of so much for so little.

In the TV series *Deadwood,* a small town transforms when the capitalist George Hearst arrives "fresh from the Anaconda," understood in the series (as in the historical record) as a massive and brutal copper mining operation. A complex social life, with characters and drama and particular language, flattens under Hearst's single-minded pursuit of the element he refers to simply as "the color," as if there were only one, as if the light of that one rendered the rest of life grayscale.

It's a simple thing, naming a measure. Ascribing value to a substance, pounding shine into currency. A simple and consequential thing: below ground, bodies bend to the hard fact of metal. Above ground, social life forms around its magnetic pull.

In the mining world I knew, underground was the domain of men. Even their aboveground talk came from underground or reached toward new properties on metal's imaginary horizon. The world above ground—house and home in whatever form, even the town if one existed—was provisional and temporary. It was the domain of women and children, oriented toward the (unseen, unexperienced) world below, paling against it in importance.

We lived where a mine was or might be; we left when a mine no longer paid off. We arranged our days around the rhythms of mining. When the mine was small, my mother's job was making sure the mine and the men had what they needed. She washed the mine off my father's clothes, fed the men he worked with, and in moments of desperation pleaded—rarely and mostly unsuccessfully—for more of his presence and concern to cohere around the family. I feel a kind of residual shame even noting this. To say it goes against an imperative that never had to be stated: whatever disruptions the mine required were justified.

If the mine was big, we could watch the line of pickup trucks come down the mountain at the end of the day shift. We lived with the mine's waste, a fine black sand called slag that dusted every road.

But if the mine was big enough, it paid for the swimming pool, the library, scholarships.

There is a sense in which the world above gave cover to the one below, an ostensible reason for existing. But the mine's actual reason for existing was whoever owned it, singly or collectively, and they were generally far from the scene. For them, the mine—sufficiently productive and not too expensive—was a wager, whether they were betting for or against.

* * *

The shape of mining—removing the desired substance from its place and separating it from what it clings to—has so many analogues that it's almost a meta-metaphor, attachable to countless other activities.

Among all these other things, it is a popular metaphor for writing. Writers or critics or teachers of writing speak of *mining* experience, as if we were pulling a story, a poem, an essay away from a great undifferentiated mass of reality, as if the selecting and refining that writers must do—of details, of images, of tones—is akin to pulling metal apart from rock. As if stepping into ideas, scenes, tones just beyond awareness—the searching, the seeking—is akin to a descent into a belowground darkness where you might find something that could change your fortunes.

But real-life mining leaves a hole where ore was, a disturbance. Byproduct compounds inert below the surface turn toxic in the air. The holes might or might not be filled in, but the process yields aftermaths.

When my mother sent me those two letters after Clementine's birth, I closed the message without replying. I downloaded the attachments and read the letters out loud to Kevin when he got home and asked him what he thought.

"Cinematic," he said, through a snack. "She's a good writer. Which I guess is not surprising."

I couldn't remember the events of the first letter, and the second came from a time on the margins of my recall. Her account departed in some ways from what she'd told me about those years, suggesting that her surrender to circumstances was not as easy, not as fated, as she had often made it seem. Some part of her, at least early on, carried a flickering awareness that things might have been different.

Years later, with two kids and a deepening sense of the ways in which the story might be the key to a lock I didn't yet know the location of, I asked my mother if I could read the other correspondence she'd mentioned in the months after Clementine's birth. In an act of incredible generosity, she handed over from deep storage scanned letters and boxes of photos and one old book—unconditionally, for me to make of what I could.

Without knowing where I was going, I went through them and started making notes. Reading and rereading, I notice what she notes and names in the way she narrates her life, and what goes unsaid or unexamined. I notice her excavating experience and selecting and arranging. I notice the sieve she applies to experience, I imagine all the mining-like operations of self-craft.

I comb through her language, skipping past what seems inconsequential in search of the valuable pieces. I recognize images, memories, situations—tangled things—coming to the surface. I search for the sensation of *something there.*

In these actions, I can imagine that I too have become, possibly and at last, the miner. Like my father did, I get up very early. I wander first to the ritual of coffee, ground the night before. I boil the water, open the filter, pour the powder into its valley. I wet and wait, wet and wait.

I can imagine a desk in my house as a worksite, where I sit in front of a screen and sift through the body of my mother's life.

It's conventional to tell students that a story has antagonists, opposing forces and opposing actors. A story has a problem and a tension and a resolution. It unfolds over time. It occurs to me that maybe this thing I'm making from my mother's words, from my life, from information—is not so much a story as it is a mine.

And if it were, I might feel masculine and purposeful comparing my work to his instead of hers. As if I could claim a workmanlike identity in pursuit of a seek-and-find narrative. As if I could skip from the limiting aboveground universe to the consequential one below.

* * *

This metaphor is available. It is also wrong.

This work may make me active, give me something to search for, make me an agent in relation to experience. It may give me a recognizable shape.

But I have come to believe this reading and seeking and finding and making sense takes place in a dimension that is, fundamentally, un-mineable—a place neither above nor below the earth's skin, but where words cross time, in a kind of mind's eye that holds the past and future, the dead and the not-yet-living. I may find my mother here, as she was when I was learning *mother*, the word and its smell and its meaning. I may find other versions of her too, and other entities that mothered me altogether.

Here, the aboveground and belowground divide no longer holds. She and I can be distinct and entangled. We can name our constraints without shame, without smoothing over. We can trace the ways we've had to harden, the ways we can still move. We can find the forces we exert, the places where we shine.

In this dimension where I descend with my coffee every morning, our stories need not be the same in order to be true. What I find here, I'm not taking. I am traveling the unseen places where connective tissue forms. It's a place to breathe with hard problems. It's a place where we can rewrite *ideas of us* that keep us from changing in the ways we need to.

II

THE USES & MISUSES OF DISTANCE

When Clementine was born and in the months after, what I wanted was to be both a mother and a person—in fact, a specific person, myself. On the level of ideas, the conflict between mother and person is no conflict at all. You are who you are, and mother is one of the many things you do. But on the level of day-to-day life, when the actions of mothering and all the other imperatives leave little room for any person to feel whole, the conflict can be immense and immovable—and even more so for someone with fundamental questions about what it means, what it looks like, what it feels like, to hang together as a self in the first place.

In the months after Clementine was born and before I went back to work, I was seasick with closeness and longed for perspective, wisdom, instruction—a manual strange enough to fit my specific experience.

> During the first months of the baby's life, you learn what it is to live tethered to a being developing so rapidly she seems to be hurtling through time, a creature cycling through more changes in a day than you do in a week. She's ready to eat too soon, ready to wake too soon, in need of fresh clothes just after you gentled her into the last ones. She's tired of it all, whatever it is, almost immediately.
>
> She wants your milk, she wants your time, and she's growing so quickly that the person she is today—this particular constellation of capabilities, preferences, and expressions; the tint of the cheek, the shade of the hair, the light in the eye—will

> be passing away a week from now, and there is no way to pay enough attention, no way with your one body to keep up, and this state of perma-change cannot help but shift your understanding of human lives generally, which come to seem like just-slower versions of this ever-passing, and all this thinking feels like a problem you would have been able to work through before, something you could've gotten past, but there is no getting past anything now when all the bits of language that have served as your survival amulets, for example *I can look after myself, I'll just get up early and get some work done, I'll just put in a little more effort, I'll be okay, It will pay off eventually* all sound hollow in the mouth and the mind because resolutions, even so small, suggest that you have some control over your body and your time.
>
> You can say them. You can write them. But you do not have your shit together: intimacy with the baby has rendered you vulnerable—

And then, in September, when Clementine was four months old, I left. Every morning I packed up and traveled south and east along the hem of Portland's outskirts, an industrial and postindustrial string of man-world enterprises grown up on the Columbia slough—wastewater treatment plant, sugar factory, strip club, strip club, strip club, warehouse, steel company shuttered lounge—to an outpost of the large community college system where I had for the first time in years a full-time job with insurance that would cover family. A full-time *temporary* job: good for a year and maybe longer if I did it well enough.

In the outside world, I was suddenly visible and unsure of how to dress myself. I did not know when I signed the rental agreement on the breast pump that it had to be transported in an absurdly large, strangely shaped black hardcase. Along with papers, books, and my laptop, I lugged it to meetings and trainings and orientations on the college's more central campuses, each with its own lactation setup, generally a single room in a distant building, un-reservable, serving potentially hundreds of lactating people.

"Do you teach music? Is that a French horn?" a colleague asked as I wrestled it between tightly packed chairs at a conference table.

In the glow of slide shows about active shooter protocols, I thought about what Clementine might be doing in her crib at Kevin's mom's house—in her car seat riding along on her grandma's errands, drinking from a bottle I didn't prepare. Did I *need* this job? Without question, we needed the paycheck and health insurance. That was simple and true. It was also true that I found myself leaning into it a good deal more than I had to, drawing energy, at least at first, from having a little bit of space.

The hurt wasn't as simple as wanting to be home and not being able to.

How do you prove yourself at an open-access college newly committed to growing the very small percentage of students who stay to get a degree? Not by research, unless it's into teaching, not through achievement of your own, but with extraordinary responsiveness to students' encompassing needs and to the agendas of your surveilling supervisors, and this requires conscientiousness, vigilance, and the peculiar skill of flowing, like water, into whatever shape the situation demands.

And how, then, do you let the baby know that you belong to her entirely when you see her mostly on weekends? By promising that she'll never see you working at home, by vowing to meet the eyes that love you with love instead of turning them toward a screen. And then how do you do all the work that doesn't fit into the fifty hours you're away? You do it after she's gone to bed and before she wakes up and during her school and during her naps, and the house you used to inhabit, the place where you used to notice and breathe, the place where you used to sip coffee and plan meals and invite other people and cut your dahlias and prune your roses and linger in the scent of your honeysuckle, becomes a racing heartbeat, a prickle on your skin, the place where you're never enough, and the office, the actual source of the uncontrollable workflow, becomes the place where you can, for a moment, relax because at least there you're in a place meant for you to do the things you're supposed to do out of the sight of the eyes that love you. There you can perhaps direct your energy to obtaining a Secure Status, which is good for your child in the long run. This is what you tell yourself.

There were two people in these temporary jobs, the other a pregnant woman due to give birth that year. You see her seldom, but she and you both keep track of who is doing what, who is showing up where. You try to make sure the surveilling decision-makers know that you can do this *even though*. You want them to believe you're interchangeable with the previous version of yourself who didn't have the baby. You yourself want to believe that.

In the mornings you lift Clementine from her crib, feed her, hold her, ready her for her ride. She goes easily, clinging to her father like a comfortable primate, eyes up and watching the world as he strides her to the car on his long legs, baby in one arm, leaning on a cane with the other.

He takes her to his mom's every morning and returns with her every evening, tacking an extra hour each way onto his commute, which was not short to begin with. Still, he often has dinner on the stove—Clementine looking on beside him in a high chair—by the time you walk in late, nipples aching if you haven't managed to pump enough, your body swollen, your mind painfully alert.

In the evening, you bounce her on your knee, her torso now capable of holding her upright, as she watches bites of food travel from your plate to your mouth. You nurse her again, wrap her tightly, place the pacifier between her lips and lay her in the crib, where she turns her head toward the wall and closes her eyes like a tiny expert on drifting into dreams.

When she's awake, in brief flashes of intense activity, she nestles her head into your neck, babbles with inflection and expression, raising her eyebrows and squinting slyly, practicing for conversation, and the certainty of her delight and interest pierces you, her memory of your body: you are still—you always will be—her home, and that place that is supposed to be respite and refuge can't be anything but tired and jagged and fugitive because that is how you are now.

The hurt was not simple. It was wanting to *be* anywhere and not being able to.

Mom visited often during Clementine's first year. She looked after things in my house like she'd looked after things anywhere we lived: kept up with the laundry, made grocery runs, cooked dinner, read to the baby (the same beloved book more times than I ever had the patience for), walked the dog, made a rhythm to the day. Her presence, and more so the following absence, suggested the possible centrality of this whole domestic world—as long as there's a person like her in the house, someone whose only job is tending to the materiality of things.

One night I walked in the door ten hours after leaving, dropped my schoolbag and giant breast pump, scooped Clementine up from the blanket where she was sitting and touched my cheek to hers as she squeaked "Hieee!"

Mom smiled up at us from the couch. The laundry basket held neat stacks of burp rags and neat bundles of baby socks I would not have bothered to fold. Her day with Clementine, I knew, was a series of deliberate occasions—intentional meals, small outings, little projects. There were flowers on the table and a complete meal simmering on the stove.

She made the home that I myself needed, the one I used to make for myself and couldn't make for anyone anymore. She curated and inhabited a spatial and temporal sufficiency that could make me feel on the days when she was there like things were all right in some fundamental way. A small glass of wine sat on the end table beside her. A notebook and pen rested on her lap.

"I don't think I could have done what you're doing," she said as her expression went soft.

Why did these words inflate a balloon of inadequacy in my chest? The balloon expanded right up to my throat, where words might have deflated it, but there were none.

Something about me was wrong. Unnatural. Wrong in my body, wrong in the hardness of my edge, wrong in my failure to keep faith in the work of tending.

She was talking about herself, not me. But I heard a suggestion. She believed I had some other choice, the one she often took: turning away from what's hard, finding a form of what might at first seem like radical simplicity in a barebones setting.

The place she would seek out is not one you can work toward, fight for, or slowly accumulate. It's a place inside yourself reachable only as the world falls away. She swerves from one kind of difficulty, but I knew, having grown up with her, that after swerving, you might look around and find yourself in the middle of other problems.

• • •

In 1975, after a simple wedding in Portland and a two-day honey-moon in Cannon Beach, my parents flew to Anchorage and met a bush pilot who carried them through the Wrangell Mountains to an empty mining camp at the foot of the Nabesna glacier. He tipped a wing toward a tiny landing strip beside a river and said, "There it is."

The small company they worked for owned several claims in the area in partnership with other investors and had to spend a certain amount annually on assessments and improvements to maintain possession. My parents were getting paid not very much to clean up the camp, get a broken-down bulldozer running, and walk it out. They drilled nothing that season.

He had a new degree in mining engineering and had worked for the same company over the previous summers. She'd applied to major in geology, but the professor who interviewed her scoffed at the idea that a woman knew enough about the work to correctly judge whether she'd be happy doing it, and so she abandoned the thought and planned to become a teacher after this adventure she was on.

They brought a few clothes, sleeping bags, a guitar, their dog Toby, a one-way radio, a camping stove the size of a beer can, a cribbage board and deck of cards, a hunting rifle, and a sawed-off shotgun for bears. He had a pair of good boots; she didn't.

After the pilot left them at the campsite, they shoveled snow out of the tent frames and found the canvas they'd use for walls and a roof. They draped plastic sheeting over the tents to keep out the weather. There was one sleeping tent and one mess hall tent left over from summers before when men in camp, my father among them, drilled samples to see how rich the property was. Whatever place they came to, then and after, my parents were walking into the leavings of previous miners. They were at the end of a long chain of people who came and went. They picked up useful items their predecessors had abandoned, including the cast-iron frying pan that's sitting on my stove right now.

The hill they camped near was rust-toned where it had weathered, water reacting with the minerals it held—mainly copper and molybdenum, but also silver and gold. The pilot was called Floyd, and he brought them food from time to time and messages from their employer. He owned the only store in the town of Northway, ran a bush pilot service, and profited in some way from every interaction my parents had with him. My parents depended on him completely. He was used to supplying a whole camp full of men instead of just two people, so he brought and charged their employers for more food than my parents could possibly eat. In the afternoons they went for runs to work off the excess. They saw wolves. They were many miles from the nearest road.

Their first job—maybe their main job—was repairing and moving a bulldozer, the Cat. Once they had it running, they were supposed to walk with it overland to a highway, where someone could meet them with a flatbed truck, a journey of multiple days.

As they set off, she rode on the engine with the dog. Their path took them across the Nabesna River during the melting season called breakup. She watched from one side while he drove the Cat through water up to his feet and worked it up a steep ledge on the other side, building a way out of the creek so that he could cross back and get her. She watched the bulldozer tilt closer and closer to vertical before the front end lurched over the edge of the bank to safety. She imagined herself and Toby stranded at the river's edge, days from anyone, if the Cat flipped.

The map showed another body of icy water they'd have to cross, and because it was called a creek and not a river, they thought it would be small. It was not. The creek was ice, thick in some places and thin in others, with channels they couldn't make out. They couldn't tell how deep it was.

Toby ran out ahead of them and they heard the cracking as his back legs broke through, felt his desperation as his front paws scratched without purchase and his body sank farther. They dropped what they were carrying. Dad flattened his body against the cold, slick surface, and inched toward Toby, hoping he'd distributed his weight enough to avoid going through himself. He grabbed the scruff of Toby's neck and hauled him over the edge and out of the dark water.

They returned to the bank and decided they had to head back to camp and wait until breakup was over—the rivers melted, the rush of water slowed—to make the trip.

With the Cat-moving project on hold, the company moved my parents to another of its properties. Floyd flew them to an airstrip five miles away from their next camp. They packed their gear through a long bog and learned that they had neighbors, V. and H., a Native woman and a white man, another six miles away. V. and H. were outfitters for flown-in sport hunters. The area was called Horsfeld because horses could graze there until the hunters needed them.

My parents saw V. and H. seldom, but from time to time they would ride in, drink coffee with my parents, and talk, about weather, about Floyd. Floyd was from Montana. Well before Alaska's statehood, he set out to go north and own a town, become a wealthy man. In Northway, he sold to everyone who came in and out. He had an antagonistic relationship with the Tanana men in Northway, whose livelihood was threatened, I can imagine, by Floyd's monopoly on supplies and de facto control over tourism and resource extraction.

Just after they arrived in Horsfeld, Floyd flew back to drop a note: the company needed the Cat elsewhere. They were supposed to go back to Orange Hill to get it. This time my mother opted to stay—on her own for what could be weeks in camp.

They had seen bear sign, tracks and scat, and he showed her how to climb up on top of a shed and use the rifle. They put their fresh meat in a garbage can and suspended it on a rope in the middle of a creek.

But in those days on her own, she couldn't eat much and didn't sleep well. It was the midnight sun, half-light all night long, and she wanted to be awake at 4:00 a.m. when the radio would broadcast Caribou Clatters, a show that delivered messages to everyone in the bush. She was listening to see if he'd made it out. If he'd gotten to a phone, it would mean he'd driven the Cat to the highway, met a truck, ridden into the town of Northway, and called the radio station.

She walked around with the dog all day, not scared but alert, suspended in a heightened version of the present moment, her mind quiet even as her body sensed her vulnerability.

On one of these nights, she'd drifted into half-sleep as a small herd of horses thundered into camp. She jolted awake with the sense that someone or something was outside, waiting for her. She stepped out of the tent into a pink glow.

She had ridden horses her whole life, on equestrian teams, on trail rides, and at sleepaway camps where girls were made to fly at full gallop along winding trails in the pitch dark. Horses had taught her how to listen and respond to the wordless language of their bodies.

Her breath and theirs mingled in the cold air as they encountered each other, creature to creature. She felt extra alive, intensely *there.*

One early morning, after my father had been gone nearly a week, Caribou Clatters read the message that he'd arrived in town, and she knew it would not be long until he made it back to camp. A day or so later, she spotted him walking back up the hill, visibly dirty, with a handful of wildflowers.

* * *

They walked cross-country to their last location through thick mosquito blooms. It was called Baultoff, and it sat high up on a bench above an airstrip. They rested on a remnant patch of snow where there were fewer mosquitoes, waiting for Floyd's supply run. Foodwise, they were down to a handful of raisins. They could hear him coming from miles away. He got close enough that they could see him shake his head *no,* turn, and buzz off after a look at the landing strip, which the spring melt had washed out.

They opened a shed and found a half-bag of seasons-old biscuit mix and a can of pork and beans. They mixed the flour with water and rationed it while rebuilding the airstrip and hoping it would hold. They were desperate by the time Floyd landed a week later.

Once the airstrip was built, others wanted to touch down there, too. There was a well-known minister who profited from flying in sport hunters. He saw the hills covered in Dall sheep and a new landing spot and flew by with a plane of fully-armed passengers. Mom didn't want to host a camp full of sheep-killing men, and Dad stood on the runway holding the rifle to dissuade them from landing.

Just before my parents left Baultoff, Floyd, whom they couldn't stop, flew in hunters who shot every sheep on the hill.

My parents would try to prevent a slaughter for sport, even while their work on the runway facilitated the sheep's demise. Whatever their personal values and priorities, from the beginning, their presence on the land made them dependent on—and put them in league with—those who put profit above other considerations. Their existence on the plane of adventure was, and would continue to be, underwritten by distant capital.

One day without warning, a man from the federal government landed in a small plane. He looked around and said, "God, this is beautiful!" He flew away, his visual assessment quick and complete. His presence foretold the end of their time in Alaska.

They worked from March until September that year and planned to go back the next season, but there was no job offer. The land was on its way to becoming the Wrangell-St. Elias National Park. The park would end mining within its borders and begin a long legal fight over the value of the metal, unmined.

That job was done, but the frame of my parents' partnership was forming. He would find work and she would follow him into places where other mining wives didn't go. She'd spend long periods of time in a particular kind of waiting in which you halfway lose sight of the awaited thing, person, or event and keep company with the waiting itself, the presences you find in your suspended state. She would test how much she could put up with and do without, materially first and then in other ways, as not-needing became a well-practiced skill, a kind of currency.

After their long season in Alaska, my parents flew home and hired on with a bigger company. They rented a house in North Idaho's silver district for a few months before the Company sent them to Leadville, Colorado, for an engineering job in a working mine. They bought a pickup truck (used, red) and drove high into Rockies with all their things in the back.

Leadville's elevation is over ten thousand feet, making it the third highest incorporated city in the United States. In the seventies, approximately four thousand people lived within its limits, down from a nineteenth century high of nearly fifteen thousand.

Bygone mining towns hang onto stories in which their sudden wealth makes them, for a minute, central rather than isolated, and Leadville has more stories than most. Its sites note varying degrees of connection to nineteenth-century celebrities including "The Unsinkable" Molly Brown, Doc Holliday, and Jesse James.

Home to one of the world's largest lead-zinc-silver deposits, it was a nineteenth-century town full of tales in which absurd fortunes rose and fell on the discovery and price of precious metals. Oscar Wilde once lectured on aesthetics in Leadville's ornate Tabor

Opera House, built by silver baron Horace Tabor and marked now by Google Maps.

Tabor founded Leadville in 1877 after discovering how to profit from the black sand that clogged the sluices for an earlier wave of placer miners. In his fifties, Tabor left his wife to marry a wide-eyed twentysomething divorcée who went by Baby Doe. She spent $15,000 on a christening outfit for their daughter Lily, who was followed by another girl named Silver Dollar. Not long after Silver's birth, the Tabors were ruined by a steep drop in silver prices. Horace died young, and Baby Doe and her daughters lived in a flimsy shack on a mine site as she tried to work the Matchless herself, got religion, repented her life of luxury, wrote bundles of "Dreams and Visions," and eventually froze to death alone, after her daughters had grown up and moved away. It was a place where the possibility of precious metal, of having it and then losing it and trying to get it back, staking lives to its price—led to all kinds of high-altitude intensity.

The wooden buildings on the Matchless property still stood in 1977, and in painting the scene, my mother made the boards blue-brown, composing in watercolor as she would in writing.

After an unsuccessful search for rental housing in the tiny boom town, my parents were allowed to use a house the company owned. It was a flaking Victorian in the ASARCO smelter yard. But when my parents first got to Leadville, they lived in the house only part-time. He was working on a remonumentation project on a mountainside above the town of Fairplay, reinscribing his employer's legal possession of each claim by surveying the property lines and writing the coordinates on a particular kind of post that he drove into each corner.

They stayed near the antique Dauntless mine in an old assayer's cabin without plumbing or electricity, or else in another tent. She was growing adept at housekeeping in minimal circumstances, taking pride in her ability to subsist—and even host others—with little.

They came by a hand-cranked burr coffee grinder and nailed it to a tree, and she spent hours driving to and from Leadville to buy groceries and water. He built a countertop between two trees and she fed birds called camp robbers while cooking for whoever happened to be up there working. She chased parts for the mine, driving for hours to Denver and back to pick up what the men needed to keep things running.

A few times I've been with my mom when she runs into friends from the old days, one of the handful who visited from Portland or else a mining person who passed through—and what they usually remember is a fine meal my mother served them forty years ago. They remember it because it was good, and because she made it in the *middle of nowhere.*

From the "front yard" of the smelter yard house, an expanse of crushed rock, my mother wrote the first letter, chronologically, of the body of correspondence she turned over to me.

In this letter from the summer of 1977, my mother is writing to her own mother who is far away on an island in British Columbia, postdivorce, building an ocean-facing house in the middle of an unlikely late-life romance with her next-door neighbor. My mother opens the letter by imagining her recipient's scene and then her own, noting the distance between them:

> *I can see you now opening this letter on a warm breezy day—the tide's out and a few gulls are cracking oysters on the rocks. You probably have two or three islanders visiting—maybe having a little wine while they help put in the bathroom sink. Then again, all the trees may be bending east as a storm rips through and you're sitting by yourself in front of a roaring fire—we'll keep the wine in this scene and maybe add one good friend—male.*
>
> *As for me, I'm sitting in the front yard with my dogs, and my*

trees, blowing in the Leadville wind. Just some thin streamer clouds are holding the mountains together. It's so bright I have to wear a hat to shade my eyes.

She makes a small map of her life, her distance from home and family, her intimacy with immediate surroundings. She writes herself into a beautiful place, a delicate place, connected to the other distant stars in her constellation of loved ones by occasional mail. I can feel the way her separation allows her to inhabit time.

Though my parents at this time were highly mobile, vacating again and again, she writes of the possibility of a vacation to the island, about my father *dragging his feet too much.* She thinks getting away *will make him so homesick for the good life that he won't be able to stand a whole winter here.* And she writes of sadness and the slowing of time following a friend's death, in an incident I never heard my parents speak of. The life insurance policy provided by the Company covered everything but accidental death underground.

She is twelve weeks pregnant with me. She refers to me as Kody—short for Kodiak, like the bear. The booties my aunt sent won't fit, she reports my father saying, *because Kody's feet will be about 15" long when he's born (7 feet tall and ready to drive a drift with hand steel etc. etc.) I don't know what will happen if a sweet little girl comes out.*

In my parents' world, what traits would be more desirable than strength, size, and skill in the old-fashioned arts of extraction? *Driving a drift with hand steel* means pounding a handheld chisel into rock with large heavy hammer, to make a hole for dynamite. Against the actual vulnerability of a baby, they imagined a full-grown beast born for immediate feats of efficacy.

The real possibility of me, the girl who came out instead, suggested a set of problems that was coming for my mother when she would have to tend to me as well as the mine and the miners and the mining, all of our needs diverging.

I think of the formation I know takes shape, even for people who never intend it: where the father's work becomes the firmament, the unmovable ordering thing, and the child's arrival makes the mother fluid, shifting endlessly.

In the letters from Leadville during her pregnancy, there are typical small updates on her health, which was generally excellent, and my position, and some discussion of material challenges around line-drying laundry in freezing weather and a theme of complaint about Dad working too much. In the new formation, he aligned more firmly, more completely with the mine.

Still, in Leadville, my parents had access to many things they would not have again for a long time. A grocery store. Running water. A relative density of other people, generally mining people, who threw my mother a baby shower at the saloon. A nearby hospital. A geographically close, attentive doctor who, with his wife, coached her and Dad in the ways of natural childbirth and cross-country skied to the hospital on the cold, bright day I was born.

And then the voice of the first letters she writes after my birth is the voice of a mom, *my* mom—an early version of the person I still know. In the first weeks of my life she is baking bread, hand-drying laundry, feeling the sudden insufficiency of time and energy.

The winter sun cut squares of light into the house. Dad was working graveyard and I napped with him while he rested during the day.

Having received—not only from my mother but most potently from her—a tale of easy, "natural" early motherhood and discomforting generalities about things working out, I come to this artifact wanting to know if it really was that way for her. Here in the letters, she's shaping the story for her reader, a distant person for whom she would wish to be *fine,* as I have wished to be fine for her. And it's a letter, too, more composed than the ephemeral *wtf omg* text messages I send during extreme conditions. The medium means it's written with a little remove from the situation. She is shaping the story for herself.

The first letter she writes after my birth, written again (as most of them are) to her own mother, is one of news and gratitude, her mother having left not too long before. She opens by saying the letter will be short because I'm about to wake up. She has had a chance to *be alone with Jessica and learn to cope with all the chores*. My father and a coworker have been *tearing around the countryside contacting people etc.* as they look for a mining property they can mine themselves. *The both of them are like a couple of kids,* she writes. *I hope it comes off for them.*

I see the thread of her attention frayed into the shreds of time when no one needs her. She writes a little here, a little there, trying to tie the ends back together. The scarcity of time to write has entered her writing. What she needs and what I need are not the same.

She reports having *gotten a little over tired off and on but am doing all right.* She says she's glad her *over tired* has not rubbed off on me and made me fussy, as if her fatigue was a kind of contagion. The underlying imperative: one must not become overtired. In these slightly antique words, I can hear an inheritance from my genteel Canadian grandmother. Fatigue as something slightly shameful, something to be avoided, something you might not want to admit, as if just being tired—in a circumstance that would make any body tired—were a measure of one's failure, error, lack of self-awareness.

Overtired is very often where I live.

A person who doesn't let herself get *overtired* can put up with anything, and that extreme forbearance is one of our jobs—or so I understand. In our family it was considered weak to lose control, and she almost never did. No kind of rage could be legitimate. We're a kind of woman in whom anger is clipped before it can bloom.

And maybe if this is the case, a person turns their attention to externalities, as she did in this letter, relaying the news of my father's occupation. As I did in my swerves toward observation in my post-Clementine letter-journals.

She was thinking not of the home that she might make but the place a mine might take us next. She wrote of these movements like they were things that would happen to her, like weather, as if she had neither a stake in them nor a responsibility for them.

The letter ends abruptly, with my waking.

Days after she wrote this letter, the tiny human in the Polaroids developed pneumonia and went back to the hospital. My lungs weren't ready for the altitude. They put me inside a cold tent. Mom remembers sitting on a chair beside me, and how the mining wives brought her Scandinavian rye crackers from the co-op and a kind of herbal tea that she still drinks.

After my eventual release my mother wrote baby exploits (smiling, toe-grabbing, paper-ripping) in parallel with the horizons forming in her mind and my father's. Her attention is *right here* on me and the house chores and then it's *out there*, in an imagined beyond.

Dad and a partner continued getting ready to split off from their employer and mine their own claim. The company was squeezing them into positions they didn't want to be in. My mother writes of these positions as if they were physical—specific jobs, specific shifts, responsibilities that were nothing but a headache—and maybe interpersonal too. With metals prices soaring, they felt like they had enough experience to make a go of it with a company consisting of two men and a few machines.

What they didn't have was a property or funds to start up. While they looked for possibilities and talked with potential investors, she reports that he was also considering more conventional exits from underground, like working for a mining equipment company or using his engineering background in some other field.

In a letter written when I'm six months old, she writes about painting—for the first time and only time in all the letters I've read. It shows up as important when she stops having time for it:

> *The art is coming alright but I don't have the time I would like to have. I have to have everything all set up before J. takes her nap and then dive in for an hour or so. I don't have enough time just to sit and think.*
>
> *I'm rushing around trying to get my wash on the line and all cleaned up before J. wakes. It's so beautiful. It was 47° by 8 o'clock and that feels like about 75° used to in the lowlands. Geo's working today. Will hope to call tomorrow.*
>
> *We are thinking about you always and need one of Grandma's inspiring letters and some of her dreams for a free, unfettered future for us all.*

Whether there's a mine in any particular place at any particular time depends a great deal on the price of metal; companies and investors calculate whether the extractable ore will be worth the cost, in dollars, of getting it out. At the beginning of the 1970s, gold was $37 an ounce; by the end of the decade, it was approaching $600. Between 1978 and 1979, the price of silver increased 438%. A change in the value of metal shifted the equation.

The three of us left Leadville nine months after I was born for a desert property Dad found with help from the owner of the company he worked for in Alaska. In a yard-sale baby buggy, I bounced over the crushed rock that paved the places we lived, then and later. Unaware of the impending change, dressed in a yellow outfit knitted by one of the mining wives, I smiled at the camera on a last buggy ride through the smelter yard.

Dad and his partner had found their property in the desert. Their main investor was a friend of the family, Jack Walker. He and my maternal grandfather were medical school classmates, and he was not yet retired, nor had he started a second marriage with a wife my mother's age, as her own father had. Both of my parents, in different ways, had grown up around his family. As their sole investor, he expressed concern that Dad's partner had no skin in the game, but he wrote the check anyway.

In the last letter from Leadville, Mom is gathering items for a trip to the post office: pictures of me for relatives, financial information for Jack. *A million questions are starting to pop up in our minds. To incorporate or not, etc. We have the option of stockpiling most of the ore until January.* In this moment of transition and risk, she is *we* and not an *I*.

My parents imagined the property was rich in silver. They imagined the mine would yield enough that maybe he'd be done working for companies. Leaving Leadville, they had to get a new pickup, having used the old one so hard it had broken right down the middle, headlights drifting apart.

• • •

The college where I worked during Clementine's first year was in East Portland, a part of town that sits on a plane of alluvial silt deposited when prehistoric floodwaters flowed from Montana to the ocean. The campus lies on the border between inner Portland and a deep, mostly flat grid nicknamed "The Numbers," where the streets run from 82nd high into the hundreds. It wasn't even technically a campus then, but a center, meaning it had few of the services the college offered elsewhere, few full-time faculty, few staff. Everyone who worked there seemed to be doing several jobs at once.

In the coming years it would become a full campus, and we'd sit listening to presentations about what it meant to work where we did: if you split Multnomah County in half exactly at our location, we were told, the west half would be one of the wealthiest in the country and the east half one of the poorest. The newspaper would begin to cover the history of East Portland, in which connected Westsiders first created density quotas for the city and then refused to increase density in their own neighborhoods, foisting it onto East Portland, an annexed part of the city that lacked nearly all of Portland's trademarked amenities, planning, and infrastructure. In the coming years the city's official publications would start acknowledging East Portland as the home of both refugee communities and longtime Black Portlanders displaced by gentrification from inner North Portland. Official voices would admit the need for public investment and representation. There would be placemaking workshops, speakers, performances, frameworks, initiatives, perspectives provided by people contracted to provide them. But that would all come later. When I started, we were just in it.

I wanted work to be the firmament, as it had been for my father, but little about what I did there could be adequately measured. The job defied the mechanistic language, the "metrics" we were supposed to apply to it. I could embody little of the efficacy I thought would save me from my state of painful shapelessness. The classrooms were still swollen in the aftermath of the Great Recession, and they flared with the students' insight and originality and contracted with their well-earned apathy and resistance. I resolved again and again to pay them back with as much love and honesty as I could access.

I spent most of my day in the monitor's glare, writing lectures, lessons, assignments that might reach them. I tried to respond to each individual's thinking-on-paper in a way that honored them and fulfilled my poorly defined obligation to hold them to some kind of standard. What they often needed, I couldn't give them: a place to stay, a place to study, someone to watch their kids (who did sometimes come to class).

I taught a pleasant young man, the father of a four-month-old, who wrote about his vows to live a new life and leave behind certain people and associations for the sake of his child. He showed me letters that he was writing to these people and associations explaining to them that he had a daughter, and with love and respect, he had to change his life. He came to class off and on for about four weeks, and then he was killed on his doorstep in the middle of the night.

Among my students there were women with three children and no partner who were living in a car. There was a witty nineteen-year-old just off meth with a beautiful infant daughter who missed too much class because she was afraid to leave the baby with the relapsing father.

Students flowed into and away from college, and unless they happened to tell you, unless you happened to see them on the news, there was no knowing where they went. There were no announcements, no events, no gatherings, no public grief, no place even to name it.

I failed to reliably remember each tiny component of my breast pump. I got mastitis twice and didn't call in sick.

I was enchanted almost daily by some wondrous delightful, energy in Clementine, as if she were a year-round summer-night sparkler. But aside from blunt sensations, aside from the occasional mileposts—I would later remember almost nothing in particular from the week, then the month, then the year.

Was I still writing, people asked? Did I find the time? Somehow, I did. Secretly, in time that was meant for something else. The work was unsettled; my poems were about losing the words with which to make poems. The neat language that felt like art to me before my life became messy, no longer worked. I had lost track of the scene or story that could anchor me to any moment.

Many of the poems I could write during this time were set in the car commute, in moments of heading elsewhere. Because of Clementine, because of my students, I couldn't even pretend to disappear. And because I couldn't disappear, I was learning to start writing from my actual self and circumstances, even if the self was relentlessly and painfully amorphous, the circumstances incoherent.

• • •

In pursuit of a *free, unfettered future* my parents pitched their tent outside of Slick Rock, Colorado, an unincorporated community in remote San Miguel County on a curved piece of land that maps drawn for mining call the Uravan Mineral Belt. At six thousand feet, it was the lowlands compared with Leadville—sandstone and sage, cut by canyons and layered with carbonized plant matter, dinosaur bones, and undulating ore bodies. Uranium and vanadium mines and mills had been producing there for decades, run by bigger companies and smaller, worked by men, often Navajo, camping with and without their families. The uranium miners dug ore-bearing sandstone out of formations called rolls, sometimes leaving wedge-shaped cavities in the desert.

My parents' dogs chased lizards up short, twisted trees that rose from the reddish dirt. I toddled from the tent. Surprised by spines, I learned the word *prickly pear.*

My mother speaks of this now as a bright time when we lived easily with very little. But her letters show flashes of disappointment, longing, and impatience that her memory has smudged away. The place became better in hindsight because the camps that followed it were a good deal worse.

Reading the Slick Rock letters, I want to see how far we were from anywhere. Google Maps shows snow-dusted sage, a drinking establishment a ways off called Dibler Cabin, and a ghost town named Tomboy outside of Telluride, the county seat. I search for the distance, over today's roads, from the camp to anywhere else: one hour and forty minutes to Telluride, one hour to Cortez, fifty minutes to Monticello, Utah.

In the late '70s Slick Rock had a post office operated by a woman whose name, Ione Rose, my mother still mentions when telling stories from that time. She might have been the only regular non-mining person in our lives then. Now the post office is in Egnar, "range" spelled backward.

Slick Rock's uranium fueled the Cold War and pre-Chernobyl power plants, but my parents came for silver, part of a wave of small-time miners returning to forgotten mines and mining districts in search of suddenly expensive metals, whose price ran high as speculators held them as a hedge against inflation.

In a world where most people have no choice but to exchange labor for a wage, where bankers and traders and executives accrue staggering riches on complex and intangible financial wagers, where young adults plan careers around the idea of skill-based worth, there's a fascination with precious metals mining, even among people who associate the activity with environmental damage.

Eyes brighten at the idea of underground wealth and the particular machinery of removal—at the idea that a life-changing fortune could be pulled from the ground, that the logic of getting it could be mechanical. My children's most treasured game is one in which they tunnel obsessively into virtual mine shafts in search of material they can trade or use to build strange homes under the earth.

My parents were living in the realm of possibility, gazing out from a high-and-outside vantage point, with schemes fanning out all over the West. In the first letter from Slick Rock, my mother wrote *we are rich again*, but what they had were *prospects*, not money in the bank, nor certain metal in the ground.

The letter is full of the doings of men and mines, joint ventures, planned acquisitions in the desert and elsewhere, who will work which mine, the details of ore samples and grades, quasi-plans about what they'll do once the mine pays off. Their idea was to make enough to stake them in a larger project—maybe housing construction, maybe more mining. They were talking about living in Vancouver, B.C. as a base of operations. *But as always,* she writes, *nothing is certain.*

She closes with a scene of the two of us alone in good weather.

> *The desert has been kind to us this past week. We had a big rain which killed the little biting gnats that were eating us alive. Since our big rain it has been partially cloudy and between 88–95—a livable temperature compared to the previous month. Lots of lovely big thunderstorms in the evening.*
>
> *Right now I am sitting on the porch while the cool breezes and blurry clouds float around—no clothes. All the men who usually are around have gone. John is in Texas for the weekend and Harry is off looking at properties. He's been gone for two weeks while George finishes up the drilling alone.*
>
> *We have to get into town, so I'll finish up. We sure are looking forward to seeing you soon. I don't know when or how.*

I would have been about eighteen months old. On many of the Slick Rock letters, including this one, my toddler-scrawl covers my mother's handwriting to the point where it's hard to read her lines. What was I trying to depict in the marks that cover this page? My mother writes the first world I saw: one in which working men came and went, making things happen. They were the people around whom our day revolved. For the sake of their activities we encamped and decamped and had what fun we could along the way. They were the people who did things. Mom and I were accessory people whose job was simply persisting in strange conditions, keeping things going, not getting in the way.

We were in a paradox of intimacy with each other and disconnection from so much else. Even when my mother reached out to others in writing, her words often positioned us at a distance, a spatial and temporal gap that can't be bridged. *I don't know when or how.*

Besides writing to her mother, my mom sometimes wrote to Bev Walker, the wife of her investor, but also a trusted maternal figure, the mother of friends she'd grown up with when their fathers were in medical school together and after.

In a neat, unbroken letter to Bev, my mother narrates a different self than in the letters to my grandmother: coherent and steady, comfortably secondary, an untroubled observer. It is as if she is seeing herself from an old vantage point, out in the desert giving aid to a big and promising enterprise.

She's just put me down for a nap, stoked the fire, and poured herself a cup of coffee. This is a serene moment reserved for writing rather than a fragment of time she's grabbed between all her other chores. She writes of their location as distant, beautiful, nameless:

> *We are so far from Slick Rock that I can't really think of that as our place of abode. There really is no name for where we are—we are caught in between Desperation Point, Disappointment Valley, and Poverty Flats—and that's no joke—those are really the names of the surrounding area. Lord help us. But we are going to be the little golden nugget amidst all those depressing places.*
>
> *I love it here. It's beautiful and strange and lonely. Sage, Juniper, and cactus are arranged like some ancient garden over the rocks and cliffs. The wind seems to be as much an inhabitant of the area as the ravens and pinyon jays—all harsh voiced—all a perfect complement to this haunting place. It all seems timeless.*

She writes herself as alone-but-not-lonely before detailing the richness of the land, which is *hot with Uranium* and packed with drill rigs. She then focuses in an extended passage on their own operation, what the men are doing, troubles with the loader, radon tests, mine safety inspections. *They are blasting today but there are still a few things left to get set up before they can really roar and bore,* she writes, having borrowed the men's language, reporting not so much from her own knowledge and perspective—she would have heard these things from my dad rather than being directly involved herself—but from his.

In this letter I can see her telling herself a story about where and who she is. It's an American story and a settler-Western story: the land is rich and empty and she is pleasantly, beautifully alone. I can see her telling herself this story even as the surrounding place names point to histories that should puncture such optimism, even as the density of drill rigs indicate that my parents were unquestionably part of something larger than themselves, not necessarily something they'd have wanted to be associated with.

With just a quick flurry of typing, I find out that there were encampments all through the district, suggesting a certain selectivity in my mother's understanding of how alone we were. I read about the wives and children of Navajo uranium workers returning in present time to the places where their families lived, remembering how the women made houses from whatever they could find, remembering how children playing in contaminated pools that formed when rainwater ran over mine waste, remembering children drowning in the same river where we swam, pointing to the government's knowing lack of compensation for long exposure to heavy radiation and early death.

My mother rose at 5:30 to work on breakfast, drove to town for groceries or parts in a trip that would take at least half a day, served lunch to whoever was working, heated water for dishes and baths, chopped wood, baked bread, and generally maintained camp. She anticipated having to make powder caps when the blasting started.

She writes about me last, my independence, my determination, my forming language:

> *She wakes up in the morning and as soon as the door opens she's headed out for her day's adventures. She eats and drinks on her own—she has her own little chair and table and accepts help from no one.*
>
> *Jessie and I have been conversing this morning on all her favorite subjects—honey, the moon, buckets, deers, and hugs. Those are her newest words. She's at that point where she repeats everything you say and then incorporates one or two words every day. She now refers to getting dressed as "a bitch."*

At eighteen months, I'm blending the language of domesticity and my mother with the language of men and the mine.

Mining has so many words for edges, for sorting what is part of what. The face; the contact; the horizon. Again, the selvedge: the edge of self. It is true that no one can know for certain everything they're part of when they're part of it. But in my mother's writing, I can see distance and also distancing.

For a long time, I thought I could avoid telling myself stories about who or where I was: better to disappear than risk self-delusion about who and where you are. But self-narration doesn't work like that. You can distance yourself from an old or bad or outgrown story. But the story stays until there's another.

As I acquired language, my parents came up short in the desert. The deposits they thought were there turned out to be a little bit of mineralization, not ore but dust. They still owed Jack, and so their small venture took on drilling contracts, hoping the machines would hold up, until they had enough cash to pay back their investor. For several months they stayed in Slick Rock, but without the comfort of an imaginary state of grace in which they were poised to prosper.

My mom continued describing our daily motions, our swims in the river, our trips to town and back for water. She complained about bugs and changes in the weather and referred to their operation as Sisyphus Mining, *still rolling rocks*. She longed to be somewhere else, longed for a vacation, longed to be heading out to someplace better. The early idyll morphed into another piece of making-do.

In the last letter from this location—this dislocation—they're looking to move on. The few people I know of, the plants and animals, the other miners and Ione, will become people I no longer know. As soon as I named the world, I lost all the pieces of it.

> *We are just on our way down to the post office. The river is great swimming these days. Jessie loves it. We are impatiently awaiting a letter from you. We need some light and laughable [Island] exploits. Jessie asks Ione each time we stop at the P.O. if there's a letter from Grandma there.*
>
> *H. is trying to talk George into building homes on BC Islands this winter, buy pieces of property. Build on them and sell. They both sit in our little patch of shade, dirty and dusty from the drill, slapping the gnats and no-see-ums and dream away but they are both serious. Maybe we are poorer but no wiser after all. George is hoping to get a few more drilling contracts to fill in till Sept.—and get us some more capital. Then maybe we'll think of heading North.*

She says now that they didn't make it out with enough cash to invest in any new venture.

Once I asked my mother when she started feeling like she'd had enough of this. "Because I can see it sometimes, in the letters."

"When we were moving out of Slick Rock, there were some things that wouldn't fit in the pickup, so your dad left a few of our things—our only set of sheets, a Hudson Bay blanket, I don't remember what all it was, but they were wedding presents and all we had—down an old mine because he couldn't think of any other place, and when he went back to get them they'd all been wrecked by moisture. Or mice. He lost some things that were important to him, too. I remember being enormously sad about that."

The adventure was turning into an ordeal, the things with which to make home eaten by a mine.

• • •

Toward the end of Clementine's first year, a week after the end of a baroque formal hiring process in which I competed for the job I'd already been doing, I triumphed at work. In the aisle of an unfamiliar fancy grocery store where I'd reluctantly stopped for a handful of necessities, I answered a call from the Powers, who told me the permanent, tenure-track job was mine. I plucked a bottle of prosecco from the endcap display and threw a tub of fresh pasta into my basket. I would be able to consider returning to that fancy grocery store more often. Almost eight years after arriving, I would take my place among Portlanders with Secure Statuses. It meant the end of piecing together gigs, of self-employment and strange tax bills, of sometimes being at home on weekdays. At a college where tenure is all but guaranteed, it meant I had not just a lifetime job, but like my father and grandfathers, a profession.

After I drove home and shared the news, Kevin and I sat on the front porch. We considered our good fortune: the winning of one of two positions for which, I was told, five hundred people applied. It seemed like employment by lottery. The other woman with an Insecure Status, who had been pregnant during the temporary year, did not win. Her child was born with a congenital, life-threatening disease, and maybe, I thought, she had not been able to make herself seem interchangeable with her prepregnant self. The job went to a man with a PhD from a top program, a well-reviewed novel, and an underrepresented specialty.

A warm spring: our viburnum tree flowered early and Clementine, with her two bottom teeth, abundant hair, and broad cheeks, toddled on uneven ground.

During Clementine's second year, we began to blink, awakening from the blindsided sensation of new parenthood. We tried various things to help us reclaim a space where our lives used to be. We tried to tinker our way back toward wholeness.

Kevin's hours of commuting, he admitted, were killing him. I asked around and found a nanny share for Clementine closer to work.

I asked for 8:00 a.m. classes so that I could spend the afternoon with her, not realizing how hard it would be to get out the door. We'd try to leave by 6:45, driving from St. Johns down the long hill into town, cruising along the flat river reflecting pink and black and green. When I reached the house where she'd spend her day in the care of a steady and kind woman, she would cling and cry as if she'd never recover from my absence, as if we were meant to be together always, as if I were the material substrate of her existence and my absence opened up a dark void underneath her.

Like the books said to do, I made the goodbye quick and walked back to the car, hoping for luck with the lights on the long, gradual hill toward campus, where I passed crowds of cyclists dressed for the weather and envied their slower pace. I walked nowhere. Even if I left work early, it was generally dark by the time Clementine and I got home, and besides, she wanted to walk herself rather than riding in a stroller or carrier. I lingered nowhere. I met needs and got shit done, racing through one task just as the next deadline was already upon me.

One afternoon at a department meeting in one of the campus's new buildings, a remodeled nursing home, we sat seminar-style in the dining hall with a wall of windows. I looked at the bright, cold day and let my mind wander, wanting to be out there like a commuting cyclist who doesn't have to drop off her child so many miles from home, wanting to let my body get tired and fully cold, wanting to leap off the daily path between house and cubicle, cubicle and groceries, cubicle and nanny share.

I thought of the towering linden trees in Cathedral Park and the way they smelled when they started blooming, remembered walking straight into the grove of them on one of my pre-Clementine rambles, remembered the surprise encounter with unexpected wonder, remembered the well of my own thoughts, the truth of my adult voice, not the one overtaken by mother-speak and teacher-speak.

During the early spring when Clementine was close to two, Kevin's chest seized with pain. He had a wide scar running from neck to sternum, an artifact of a thymectomy when he was younger. His bones there were wrapped together with wire and they ached sometimes when it rained, but this seemed worse. It was unbearable when he lay down. On a Saturday morning after a sleepless, upright night, we drove him to the hospital, Clementine in the back babbling on and looking out the window. Kevin thought lupus pericarditis, something he'd had before more mildly, was likely. But when we arrived, he failed the initial heart attack screen and was rushed away to a cardiac unit, sweating and agonized. We had been through similar episodes before, but never as parents. Before, I would have stayed with Kevin all the way along, but now I had Clementine, and I wasn't taking her back into the guts of the hospital, where she'd hear and see and be exposed to who knows what. There was nowhere for me to be, so I took Clementine home and waited for a call.

After days of waiting for a cardiologist to have time to see him, we'd learn that it was in fact lupus pericarditis, treatable by knocking down his immune response with a heavy dose of steroids and hoping for the best.

But before that, Clementine and I lived in the long, arctic plane of waiting: something about it—her tiny body, her constant utterances, her endless attempts to make sense—inside some kind of fatherless semi-emergency invoked a long-forgotten place and time I couldn't name. The house was quiet. We let Kevin's mom take hospital duty. I did no work while Clementine was awake. I was her only person. She bounced on the couch while holding my hands.

The air was cooler and every color was brighter and sharper. Clementine's utterances were brighter and sharper. We had chips and hot chocolate for dinner. There was no reason to meet the needs of any time but now.

Did I have postpartum depression in the years after Clementine was born? I had energy spinning over a sea of wordlessness. I *wanted* to sleep for days, but I didn't. I *could* get out of bed, and I did, again and again, all night long, whether I was well or ill.

The internet assured me it was normal to be tired. But there were solutions, it said. Great! It *is* important to take *time for yourself.* Don't forget *self-care.* Take a hot bath! Make sure you're getting enough exercise! Take that time for *you, Mama!* Pedicures are your best friend. Every bit of prescriptive wisdom I could access seemed to assume that some retrievable part of me knew how to be present as a coherent self in relationship with others.

There was no prescriptive wisdom for the times when we really needed it; for example, when both of us were ill with something the baby brought home, and she was recovered and wanting to play and eighteen months old and needing constant supervision and interaction and there was no way for anyone to rest, not really, and no way to call in sick without inconveniencing all of the students in such a way that everyone would know I was not interchangeable with my pre-motherhood self, that I had this swirling need of a family, and if I called in sick every time the baby or I or Kevin was unwell, I would be working maybe 30 percent of the time.

Did I have intrusive thoughts? Maybe? Doesn't everyone? Did I have thoughts of harming myself or my child? Never. Did I find it hard to stop imagining scenarios in which my child might be harmed or left alone or made to have experiences that would force her to turn away from herself the way I kept turning away from myself? Sometimes.

I saw a therapist. Twice. Her office was in an old converted house. The second and last time, I walked through the entry and settled onto the cushion. The white sky glowed through her gauzy white curtains. She was pregnant with her second child and spoke little. She asked what brought me there, and I didn't know how to say it.

"Suddenly last year, I got a lot of things I've always wanted. The job I didn't think I could get. Financial security, finally. But."

I wake up early with the moon still in the sky and Clementine is learning to say the word moon, reaching out to touch the untouchable and it is as if she has every right to this moon and she loves this moon without self-consciousness and I have become a kind of absence. I don't know how to feel another way.

I did not say even this. I didn't have these words. Whatever I did say, the therapist did not affirm, did not question, did not paraphrase. Did not vocalize, did not nod in acknowledgement.

How to tell the silent therapist that sleep deprivation, vicarious trauma, and an excruciating cognitive load had left me unable to record a memory? That I didn't know a way out. That I wanted to be able to go home.

How to tell her the child I had been: wandering in camp, drawing, reading, imagining. Like Clementine, a little light, useless and fluid, acquiring the language of industry and frontier rather than the language of care and continuity.

With my fingertips, I combed and recombed the tassels on her couch pillows. I said a thing, she said nothing. Repeat. I was illegible, a scrawl across the page.

III

UP TO THE LEDGES

Consider the self-edge, the boundary between what you are and what you are not. Imagine the self forming moment by moment, making and unmaking itself in response to experience.

It finds its edge when it encounters something harder. Something to say, *you're not this, but that.* It must form also in failure: *I thought I could do this, but in fact I could only do that.*

Imagine a feeling of expansion when a barrier—social or physical or circumstantial—is removed. This is what I believe my mother loved—still loves—about what she would have thought of in the seventies as wilderness: removal of inhibiting structures and imperatives.

This romance of being alone: when it's just you and the trees, you can imagine yourself as larger. You can wear what makes you comfortable; if it's warm enough, you can go naked; your time is yours. Your questions can go quiet. You can imagine your heart is larger with little risk. It's not hard for me to sympathize with needing the world to fall away for a little while to recover yourself from your surroundings.

There's an age at which the notion of a distinct self forms, which is also roughly the age of first memories.

We lived up a very long road; we were often snowed in. I remember looking down at my sweater-covered arm in the white afternoon blizzard light and realizing I was a person. Then, the terrifying feeling of reaching back further in my memory and finding nothing, as if the consciousness I called *I* materialized right there in the middle of things.

What did it mean, there and then, to be a person like other people? Mother too close to see, father so often gone he had to be more than half imagined. People beyond our immediate family were either far away, or just around the bend in the road and dubious.

Wilderness—as opposed to a corrupt or overcrowded or over-industrial urban life, the settler fantasy of empty, untouched land, and even of being truly alone—was an ambient idea for people of my parents' time and place, so common as to escape much questioning for most people. For me, because of reading and conversation and relationships, it's a false notion in the aftermath of genocide, and offensive as an idea that enabled it and would still like to obscure it. If you look at a place with your wilderness glasses on, can you see relationships, even the ones right in your face?

A thing I know with certainty about wilderness is that it required selective vision. It was never just us and the trees.

2013

Our first night in the new house, I couldn't sleep. In the shade of trees and set far back from the streetlight, the air toned blue. Our things amassed in box islands.

The new dishwasher beeped four times in the dark, its cycle finished. I rolled over and it beeped four times again. The new fridge beeped if its tricky doors didn't seal tightly. The sounds traveled through the house like sonar pings.

Out the window, I knew, there was a ravine and a dark wall of cedar and holly, a bank in need of restoration. I knew there was a vast finished basement below me with a bathroom that would need to be cleaned and stocked with towels I didn't have. I knew that every bedroom had a wide closet. I knew the birdsong would flare just before the light.

In the old house, no place was more than five decent strides from any other place. I could clean it thoroughly in an hour. The night before, I'd slept in the old house for the last time.

It was supposed to be easier here, deep in the imagined Elsewhere of my childhood, a rambling house in the hillside Portland neighborhood where my mom grew up. Ranch average, sure, but compared with our old place, palatial. I could feel, in this new quiet, the touchless sensation of so much space, the uncomfortable feeling of floating. Inhabiting the space for the first time, I lay awake and wondered what we'd bought.

Our new house lay three miles south of downtown, separated from the city proper by an ivy-matted hill and Oregon Health & Science University. The neighborhood was suburban but not technically a suburb. Even when cars bound for the south and west suburbs clogged the freeways, traffic to the new neighborhood generally flowed. We had a city address despite the wide, forested lots.

Portland's extensive streetcar system once served our old neighborhood, carrying laborers to and from factories and taverns and soda fountains and markets. It never served the new one, which was built later. The new neighborhood rose on old dairy farms during the decades after World War II. It was shaped for families with cars, a man who went to work and a woman who stayed home and took care of things. There were no drug treatment centers, no shelters, no public housing, no duplexes, even. It was full of Mommy-Daddy-Baby houses with little country details like rooster weathervanes and ranch rail fencing around expansive yards. It was extraordinarily well served by bus routes to downtown. It seemed built by and for people who had some measure of influence.

We didn't choose it easily, and before Clementine it's nothing we would have sought.

After months of searching the blocks of Eastside bungalows and not finding the right one at the right price, we spent a day on the Westside and noticed children out in the neighborhood walking and riding bikes. We cruised by a tween girl flopped on her stomach in the grass, reading a paperback, flip-flops dangling sideways from her toes. We fantasized about Clementine encountering the patch of forest behind the house, learning its ecosystem. Though there were no sidewalks, a steady stream of pedestrians strolled on the shoulders of busy roads, headed for the farmer's market or the cafe in the '60s strip mall that marked the neighborhood's commercial center.

To get to work, I'd travel downtown and then turn east, up the long, slow grade through the condo-and-restaurant streets toward campus. But Kevin, now the manager of the campus library in the same system where I worked, was a few minutes from work and his parents.

Our new house sat in a clearing on a bank of clay soil. It was a 1950s ranch, but an unusually large one, stretched across the lot. Like a wide, old sedan, it invited you to lean back and spread out. The previous owners, a manager at the local electrical utility and a homemaker, had kept the house eerily pristine since they bought it in 1968, the basement bone dry and the tile grout perfectly intact.

When we drove up the street, we could see, through open garage doors, boxes stacked neatly on shelving systems and hand tools arranged on pegboards, rows of wrenches and pliers descending from largest to smallest. The normalcy seemed nearly exotic. We were no longer in the land of making-do.

Outside the huge picture window of the new house, we watched vintage Mercedes and Volvos, ten-year-old Subarus and Civics, and the occasional Tesla cruise up a gentle, forested slope toward the neighborhood's small "town center." And so many joggers: headlamps carrying fleet-footed bodies up our street at all hours.

Driving away from the new house one day in the spring just after moving in, I was stopped at a light where I could see a track-and-field practice, a girl with a javelin. Growing up in Idaho, we weren't allowed to throw them. Like soccer, Advanced Placement classes, and public transportation, javelins were for Elsewhere.

A javelin! A javelin meant space, infrastructure, expertise; it said you'd hand a spear to a teenager and teach her how to hurl it. A javelin whispered that my children would grow up someplace where full social participation was possible, even for girls.

I watched the girl adopt the stance, take her first steps. I watched the potential energy build across her chest, ponytail bouncing. But just at the moment when she might have launched her honed projectile into the cottonwood-flecked air, she crumbled in a self-conscious, half-giggling posture that comes with a bodily loss of nerve. She had a javelin, and she didn't even know it.

Ignoring the constraints that had defined our lives up to this point, we nurtured fantasies of who we—who our children—might become here in the land of resources.

In the new neighborhood, nudged and helped by my parents into an ostensible good life, I felt the hand of traditional homemakers, women with the time and mandate to attend to material details. The new house had a whole room for storing home-preserved food in the basement, a corner of the kitchen with a built-in desk and a phone jack. I could picture the lady of the house sitting at that desk, writing notes and recipes, paying bills, planning trips, assembling photo albums—doing all the things I thought would make a home.

Could I be a woman like this? A woman who preserved and curated and assembled, who made a net of memories and people and images for Clementine's sake? I wanted to think so, wanted to think that by buying the house I was somehow buying this kind of life.

I saw them in the grocery store wearing kerchiefs and galoshes. I saw mailboxes that said Mr. and Mrs. plus a man's full name. I saw them in the array of annuals for sale everywhere in June: plants for people who have time to arrange those bright beds and window-box planters.

Waking too early to write or waking too late after a night of interrupted sleep, I in fact did not generally make the bed I slept in.

Still, I thought I might be able to be this kind of homemaker, doing things I never would have thought of before children, the kind who curated the life of a family, maintained ties, made a place and an *us*. Despite my job—still had it, didn't see leaving it—despite my writing compulsion, which had come down to a semi-desperate transcription of the beautifully strange things Clementine said, alien things I found my own voice saying, it was my ambition in the new house to shape myself into this mother-template, an inheritance from my mother and her mother and her mother. I imagined we were buying the material underpinnings that would keep me from falling short.

But my imagination didn't account for the actual circumstances.

The new house looked out on a thick, dark wall of introduced and invasive holly and laurel trees that obscured the trunks of native fir and cedar and choked out most of the native ground plants. *We'll just take those out,* we thought.

The front yard was matted with weeds. *We'll just clear it, make beds.* The fence was falling down. *We can replace that.* In the new house, decades-old rhododendrons grew out of control. *We'll prune those up.*

Before Clementine, we'd spend weekends in the old house moving dirt and building screens and trellises. What we'd done easily in a small house with no children and unsteady work was exponentially harder in the big new house with a toddler and two consuming jobs. The rooms dwarfed our small collection of furniture. The cold clay soil out front resisted our shovels. The dense tangle of invasives kept winning against our unaccustomed hands and office-withered musculature.

The hillside was chilly, the soil was chilly, and the neighborhood was too, in ways we didn't expect, as bright and optimistic as it seemed at first. No loud parties when the baby was sleeping, but also no requests to borrow pans and sugar, like there had been in St. Johns, no divided iris bulbs on offer. No front porches.

We didn't feel much like we belonged there, and we weren't sure what it meant to belong to a place like this. If I walked into the driveway, it was because I was headed somewhere else, for some purpose that had nothing to do with here.

I kept a little flame, a little mutability, a refusal. I loved my girl: I sweetened to her sweetness, brightened to her brightness. I could not have told you why I still could not be the Mommy of the Mommy-Daddy-Baby world.

After we moved, same as before, my job bled so heavily into the off-hours that I couldn't tell off from on, family-self from writer-self from work-self. I woke up early to grade, I stayed up late to email about committee work, I just-a-minuted Clementine on Saturday morning to sit at my computer and get the next week's planning done. I went blind-tired during her phases of night wakefulness, when a terror started her screaming at exactly 12:30 a.m. and I had just shut my eyes and only four hours remained before the alarm went off.

I kept seeing horizons. *When I get tenure, it will be better. When I fix this problem with this class, it will get better. Once we finish this new curriculum, it will get better. If I can just get through this week, it will get better.* I was failing to tell myself the truth.

From my father, I had learned extreme subordination to actually unending kinds of work—the only way I could understand that I was real. From my mother, I had learned the way smoothing over fundamental fissures by reaching for familiar trappings, a familiar role.

• • •

In Eastern Oregon a crescent of gold-laden rock one hundred twenty miles long and forty miles wide stretches between the John Day and Snake rivers, forming what mining historians call a district, a particular locus where the mechanical and igneous process of mountain-making infused a place with metal. Through lode (tunneling) and placer (panning, sluicing, dredging) mining, the district has given up three and a half million ounces of gold and an equal amount of silver since the nineteenth century.

"District" suggests proximity, as if you could travel along the crescent from camp to camp, little town to town. But the roads are made to go from the main highway to each mine and back.

After Slick Rock, we moved to a just-revived gold mine in this district, outside Granite. The nearest town big enough to have a grocery store was Baker (now Baker City), then about ninety minutes from the mine over sometimes rough roads.

The mine was backed by a well-known New York financier who sent his managers occasional gifts of boxed fruit. Before leaving Slick Rock, a falling jackleg drill had cracked Dad's femur, and my parents' finances were still slightly negative. In hiring on with a company, even a small one, they were trying not to take another risk.

We might have been counted among Granite's seventeen people in the 1980 census, except that we lived a couple of miles outside of it.

Granite formed during the gold rush of the 1860s. Prospectors living on the coasts backtracked along the Oregon Trail or traveled up from California or Utah. At the peak, which didn't last long, it had five thousand residents, more than half of them Chinese laborers and re-processors who bought ostensibly mined-out claims when they weren't allowed to buy fresh ones and found ways to extract more gold from the piles of slag.

In 1939, a Work Progress Administration employee interviewed a Mrs. Neil Niven, who had been one of several schoolteachers there in the 1880s. Of the bygone town, she said:

> Granite, or Independence, as it first was called, was built in the heart of the Blue Mountains. As you know, many creeks roar down from the mountain springs into the canyons. The mountains are rough, tower high into air and flatten out into rocky, almost impassable flats at other places. This rough country presents many obstacles hard to overcome. The roughness, coupled with the unfavorable climatic conditions are, at times, almost unbearable. The Granite country could be and usually is nature at its best and worst.
>
> Each season of the year presented peculiar problems.

The revived mine had been going for a couple of years when my parents got there. A massive fan ventilated the long decline to the Face, where a small crew drilled, blasted, timbered, and mucked out a few cubic yards a day. Only a few people worked the mine, but one of them wrote and self-published a book about the experience, *Granite Stories*. Though billed as a novel, it almost reads like nonfiction, a story from life, though the author, Vance Bennett, says he took liberties with some facts and characters. In the small parts where my family's story intersects with his, names are unchanged, but his timeline isn't always right.

Bennett's general description of the place and people, though, resonates with my mother's letters and backfills information she wouldn't have known. Before my parents arrived, there was a small crew that included a couple of people with mining experience and others who drifted into Baker City between restaurant or mill jobs and were sent on an hours-long drive to the mine by the unemployment office. Though the area was historically a mining place, none of them was from there and local knowledge of it had dwindled as the mines shut down following World War II. The Granite miners learned to mine together with informal lessons from a 1917 mining handbook.

Until my dad came, they were more or less unsupervised at the mine site, working on an irregular schedule. The Company's checks didn't always cash. Bennett uses the word "romantic" to describe the general vibe. He describes the Granite miners as members of an easygoing, nonconformist generation trying to live free at the tail end of time not yet constrained by technology and professionalization.

In Bennett's story, my father appears in the mine without warning, leg in a cast from hip to foot, cursing at a too-large timber that they're all trying to shove into place. My father had been a worker since he was a kid bucking hay bales on a ranch, then pumping gas in Seattle, then working on a drill rig in the Gulf, all before his various mining jobs. But at the Cougar, he was a boss, and distrusted the moment he showed up. The Company had sent him to make the mine produce, which meant making the crew conform for the first time to shifts, schedules, and on-duty sobriety.

Bennett repeatedly mentions the superiority of our lodgings to the miners'. The Company was supposed to provide us with a place to live, while some of the miners appear to have been squatting in abandoned houses. We may have been warmer, but in Bennett's telling they seem to have had running water, while for a long time, we did not.

My mother may already have been pregnant with my brother when we got there. A year earlier my parents had been dreaming of freedom and the good life; by the fall of 1979, they were going wherever a mine was.

The material, marital, and interpersonal conditions weren't suitable, but this is where my mother started dreaming of tulips and flower boxes and patios and lawns. She imagined that the Company might build them something that looked, even on the surface, closer to what she knew of home.

The Company was supposed to build us a cabin, and until then we lived in a small trailer. I remember nothing about the inside of it, just a vague sensation of very close quarters and nowhere to go. When the weather was nice, it felt like we mostly lived outside.

I remember sitting with my mom in the yellow Datsun pickup during a thunderstorm during this time before the cabin was built. Big raindrops fell across the windshield and bloomed daytime shadows across our faces.

"Did you see that? One . . ." She started the count and then *boom*. "It's close."

I didn't know if I saw *it*, the all-around flash, no longer than a blink, something there and gone too quickly. My mother reveled quietly in the lightning show—an event big and dangerous enough to mean something later in the telling. She loved this immediate space of *what might happen*. I wanted to go back inside, back to the small animals I was arranging on the floor when she scooped me up and said we had to go.

"We have to stay out here until there's no more lightning."

I wanted to know why. In the trailer, in town, in the car, in the tent, there was so much of this waiting that my waiting self felt too heavy to carry.

"The truck has rubber tires, so if the lightening hits us nothing will happen. And the trailer is a little tin can up here on top of a mountain. Do you remember the thunder song?"

The song was about thunder, moonshine, and *the Devil's thirst*, substances and metaphysics I had no notion of, just as I had no way to make sense of lightning as electricity falling out of the sky, no way to understand what conducts electricity and what doesn't.

We sang the thunder song. I fit my words to my mother's words and caught my mother's feeling: the two of us being brave together, alive and waiting in a place that was so far from anything else.

In the letters my mother gave me, the ones written from Granite are thicker and more numerous than the ones from any other place. Compared to the desert, the Granite camp was harder to leave. The roads were rough for a long way, and the drive to town could be once every week or two—when the road was passable—rather than every other day. The letters from Granite—at least the early ones—read like she had extended stretches of time.

Stretches maybe too extended. In place of the "beautiful no-place" myth she painted in the desert, her letters from Granite read as if she's stuck in a situation she wouldn't choose, alternately needing to name it as bad and then needing to convince herself it's not that bad.

The first one's from the fall of 1979. She'd have been pregnant with my brother at this point, and we were snowed into a tiny trailer during an extended snowstorm. The plaintive quality of her correspondence begins with the address.

Dear Mom, Dear Dear Grandma,

Are you up for a tale of woe and hardship? It's hard to find much that's positive to say about our little set up here except that it has a snowy panorama of misty blue forested hills; and it has to get warmer sometime.

It probably won't be so bad once this cold snap breaks. It was 40 below our first night here and only slightly warmer last night. Jessie and I have been totally confined to this little space. The heater's blasting all day and all night while we run around in our altogether trying to keep cool. The pipes only unfreeze about 4 o'clock in the afternoon so I can do dishes, wash my face, make dinner, give Jessie a bath, and by then we've lost it again. Will we go insane after 4 weeks in a little sweat box? Poor Jessie can't even climb up on the seats because they are too high so she spends the day on her bed walking up and down it and reading her books till she sees one of the dogs out the window. Seeing the dogs out the window is a highlight.

She'd have to wait for a trip to town to bathe herself. There were problems at the mine: a pump clogged, so my dad was with the miners, waist-deep in freezing water to keep the operation going. All of us were ill. Their trip to town was delayed by a huge cave-in at the mine with no one hurt, and Dad fighting his boss—a person offsite—not to return to the cave-in area and risk a loss of life. *I think George is losing his sense of humor about this operation,* she wrote. *Too many wet socks last week and the problems seem to be overwhelming the laughability of the bungling.*

Again, she and my dad were talking about ways out of this lifestyle. At this point, he was still thinking about heading a company of his own, but after just a few months in Granite, he was understanding how difficult it would be to mine on a small scale. He thought about doing sales for a mining equipment company and training to become a cabinetmaker, maybe staking some claims on the side. These alternative futures are a drumbeat that allowed them to continue in the present. They never made a step toward any of them, and reading about them in the letters is the only way I'd have known my parents ever considered doing something else.

During this particular snow-in, we ran out of food and lived on sourdough biscuits for a few days before we could make it to Baker. When they finally made it out, they stopped at the laundromat in Sumpter, which had a coin-op shower. There, while Dad was showering and Mom was doing laundry, a young woman with an eighteen-month-old boy opened the door and invited us next door for coffee. The woman and her husband were opening a restaurant off the laundromat. My mom described it as *very old and done with lots of charm and grace.* She said they *owned a gas station and store in Corvallis and went looking for the ideal life and wound up in Sumpter. This will be their first season.*

This vignette might be the only one in the entire body of correspondence in which my mother notes an encounter with another woman and describes her favorably, as in some way like us. I sense that she was at least trying to see this couple as akin to them, as if maybe she found it affirmative that someone else with a baby would come up into these mountains. I don't know if the restaurant ever opened. There's no evidence in her letters or in memory that we ever stopped there again.

On the way home from Baker, we raced back up the mountain so that she wouldn't have to put on chains in the dark—Dad may have gone back to the mine after the stop in Sumpter. The letter ends in a peculiar and characteristic mix of relief, scenery, and deprivation. In one moment:

> *The sky is heavy with clouds this afternoon. Everything outside looks soft—the air, the light, the piles of snow. The hills holding up the slow moving clouds. Even the sounds are soft as the wind sifting through the pines.*
>
> *We are much happier this week than last. Just needed a little adjustment period.*

And then a few days later:

> *Jessie has finally gone down for a nap. She and I both caught George's cold yesterday and she has a little trouble sleeping.*
>
> *She doesn't have any chance to crawl, walk, or climb at all which is a pretty sad situation for her. She also has no change of scenery—no new room to walk into, no new toys.*

Do I remember this specific scene? I remember the feeling. It shines like a fleck of essence in the neural stream—mother and child confined, stuck in the scarcity of experiences, in a hut of sweaty waiting and grating. In all the years I didn't want children, or wanted them but feared I was unequal to the task, moments like this glinted through my mother's comforting assurances that things generally work out.

By the time the spring melt came, Mom was heavily pregnant. Her doctor—not an obstetrician but the only one available—told her that she wasn't getting enough exercise. She needed to walk more.

She tried. But spring in Granite meant mud, not just smeared on the trailer's metal grating stairs, but a wide, glistening field of it. She couldn't step outside without sinking in deep.

The promised cabin, the promised running water, were slow arriving. It's possible that the people doing the work were not trying to make it easier for the boss and his family to stay there.

Writing to Mrs. Walker, she spends less effort trying to convince herself that she's okay and writes as if she's a regular homemaker, writes as if all the circumstances aren't arrayed against her, writes as if she isn't living in a jobsite littered with the inevitable garbage of mining in places where the dump is distant and no garbage truck exists. In hardship, she latches on to a familiar, inherited story about her role and turns away from newer stories that would change it.

> *As the snow melts little by little each day I realize how much work this place needs. Our house sits in a sea of mud, mostly, except for the places where it might be nice to plant a garden. Alongside the house they have dumped black mine gravel making it hard to envision tulips and crocuses coming up. We are surrounded by lovely pine forest but it is strewn with scrap metal and wood, pipe, wire—everything imaginable from beer cans to junked cars. I can't believe the insensitivity of the people*

who have spent the last four years turning this beautiful mountain retreat into such a blight. So George and I have our work cut out for us. It all starts next week.

Next week the house will start to be finished. I agree with you about setting a deadline for these balls of fire. The weather is pretty good tho so I do have high hopes. What makes this nice is that I have at least had a little say about what the kitchen and bedroom will be like. Just little things like a few large windows will be nice. We will also have a door out of our bedroom that will lead to a back patio. The back of the house is in the trees and is private and shady. It will be lovely when we get the piles of junk cleared away.

After all that is done, house, cleaning of junk, patios built, flower boxes made, grass seed scattered (oh let me see) vegetable gardens put in (optimal) but before the baby comes, Geo is going to build a tent frame and put up our summer tent on the hill behind the house. Mom will be staying over when babe arrives and I hope everyone comes and visits so we will need a guest house.

I write all this and sound to myself so absorbed in domesticity. I think of Anne Lindbergh's book, Gift from the Sea and realize that I am really in the building, collecting, gathering stage of my life. I almost can feel myself secreting a shell to expand and hold my increasing family. George is happy in this too. He's taking care of buying all the necessities while I'm in Portland—stove, fridge, bed, cabinets, etc. I'm reading this book right now, The Women's Room. I don't recommend it. It's really well written and the author, Marilyn French, has a marvelous second novel out. But this book is the kind that makes you so angry. A women's book about our wasted little lives etc.—at least that's how it seems in this first little bit. And here I sit, fitting into the mold so well. Trying to write you a letter, play with play dough and worry about dinner all at once. But actually I'm enjoying the book because it's making me think about a lot of things I never thought of before. George, Jessie, and the people they are—our lifestyle—my future and my goals. It all seems very real, good and solid (tho not very easy) compared to the characters in this book. So I'll keep reading it so long as I can enjoy the writing and keep the propaganda in a good perspective.

Granite, small as it was, had a couple of families whom Vance Bennett as well as my parents referred to as "hillbillies." In Bennett's telling, one member had driven to Oregon from Appalachia with a sawmill in his truck, having chosen the place more or less at random. When the women of Granite, who would have been mostly from these families, came to welcome her, my mother described them in her correspondence in terms that emphasized their difference. Here she makes similar moves in a short space: experiencing hardship and reaching toward the class position she was born into rather than the one she currently occupied—embracing the idea that one could just set a deadline, and the working men would meet it. I can see it as a way of reaching for an idea that would let you have at least the illusion of control over daunting circumstances, the false comfort of a sense of superiority.

She reads second-wave feminism as a form of "propaganda" and an insult to her choices. She couldn't accept a worldview that regarded her work as meaningless or small.

In the months before Andy was born, we drove across the still snow-bleak state to Portland several times: down out of the Blue Mountain mud roads and onto I-84 and the melting fields, under the iconic green steel overpasses, from rest area to rest area, along the long Columbia, and into the tall-tree and low-cloud country. Even a small city dazzled, driving into it at night. We looped through unbelievable structures as freeways merged and split and wound our way to Grandma's house in the suburban development just across Barnes

Road from Catlin Gabel, the pastoral private school where she'd found a postdivorce job as a school librarian.

It was a long distance from the scrap-strewn mountain to her sunken modern living room. Six feet tall and elegant, Grandma gathered people easily, including a next-door family of kind intellectuals who ran an academic publishing enterprise. I floated into their perfect garden with wonder. I learned what cherry blossoms looked like in a paradise of ample and thoughtful adult attention.

So many small worlds made for children rather than working men. Visiting Catlin's Lower School, I sat under a lofted parachute with a circle of other children. The Oregon Museum of Science and Industry held a child-height table full of warm new chicks that I was too wonderstruck to ever touch. A downtown department store had an escalator, the first I'd seen, and an enormous clock that seemed to tick the city forward.

There were streets called boulevards. There were streets called avenues.

I did not understand why we had to keep leaving, why we did not just stay and make our life there rather than returning to a stretched place where good things from food to friends to water ran short and scarce.

We drove back East, stopping in Baker to meet Dad for dinner at a restaurant attached to the usual motel we stayed in when my parents were desperate for a hot shower, the tile roofed Eldorado, a place that loomed large in my imagination because of its swimming pool. I was thrilled to be reunited with Dad, and we wrestled in the motel room until late. The next day, we did our town chores and started our trip up the mountain, stopping to visit family friends on a ranch, where a calf was being born. They let me tickle its nose with a straw to get its breath started. Mom and I drove up the mountain a little ahead of Dad. She was expecting to find the cabin done, the scrap and trash cleaned up, the foundations of the home-vision becoming reality. *When I pulled up it was a bit of a shock,* she wrote to Grandma.

> *The trailer is parked right in front of the house leaving little room for our three vehicles. Garbage was strewn all over as well as piles of scrap lumber, plastic, all sorts of chains and propane bottles, broken tools. After I worked so hard to clean it up before I left I burst into tears on the spot. It looked like a combined junkyard and used car lot. Inside didn't cheer me up much. All is pretty bare and utilitarian. The neon tube lights in the kitchen have no covers on them—that kind of nonsense. Well, it's a good thing George was a half hour behind me in the Datsun. He's not too happy with things either and it's too bad he has had to work so hard trying to compensate for everything they didn't do. It was just one of those evenings. Jessie wanted to go back to your house and I was making cheese sandwiches which I forgot about and toasted them into oblivion, filling the house with smoke and killing the plastic flowers that they had put on the toaster to cheer up the room a little. So anyway that was our fifth anniversary.*

The Company had not made the homelike home she'd imagined. Still, we moved into the cabin and hoped the water would be hooked up soon. Though April first had come and gone, it was still *snowing, snowing, snowing, snowing, all week long,* she wrote. *So Jessie and I have been basically confined to the indoors.* She describes a scene in which I'm desperately trying to get our old dog to go with me on *an expedition to the North Pole.*

But soon she finds a way to recover, to prevent her momentary desperation from turning into a bigger complaint.

> *As usual, when I got everything in its place, clean and easy to work with, it all seems much better. George built a nice ta-*

ble and lots of counter and shelf space. Our fridge and stove are both clean and shiny new. The fridge is not real big but has a good sized freezer and I can't seem to get it even half full after using the little 2' by 2" trailer version for so long. The company—in their usual strange way—gave us a microwave oven for the kitchen. I would have settled for a piece of plastic over the tube lights but I got a microwave oven. But anyway, I'm feeling pretty good about the house. When I get the trailer cleaned out, moved, the front yard done I'm sure I'll feel good about everything. We may even get water soon. The big hint is that S. built and installed all our bathroom fixtures in his house so I'm hopeful that he plans to hook up the water soon.

I have just discovered that I have to bake bread today so I better get on the ball. I wish the microwave could do something useful like make bread in three minutes.

This letter was one of the two she scanned and emailed me when Clementine was tiny. It was one of those I opened that sent me on this prospecting of family history I did and didn't know, this reprocessing of things I had forgotten. When I first opened it, I saw the conflict between her background and her surroundings, her naïve and upper-middle-class assumption that The Company, or the men who built her cabin while likely living worse conditions themselves, would understand the ways she cooked, would sense her aesthetic priorities.

Maybe I side-glanced at a sense of homemaking dependent on things like bread your own hands made and pleasant lighting. Maybe I resented this home vision because the standards I inherited from her had often felt like grit in the gears of my own engine, the other imperatives humming inside me. Reproducing my mother's version of domesticity seemed barely possible without children and impossible with them. And still I can't stop wanting the kind of home that *can* be, *could* be, *might* be, among other things, a site of creativity, connection, gathering, abundance.

Maybe it was the sense of her powerlessness in the situation that made me want to distance myself from the scene.

Reading the letter again, after reading the before-story and with more years with my own children gone by, I see her—on her own, heavily pregnant, and snowed in up to the ledges in a place that defied her vision, with little control over what happened to her, such a small wheel to steer.

I want to be able to treat this past version of her like a friend.

I want to send her an absurd emoji-laden text message of encouragement, send her a little care package full of beautiful things in the mail.

I want to reach through the page and take hold of her beautiful hand.

A couple of months later, when she was in labor with my brother—whom I was still hoping would be a kitten, or if not a kitten, a sister—my mother woke me in the middle of the night to tell me the baby was coming and it was time for her to go. I was two and a half, but I remember it: she and Dad in my room, the cool tone of dark. She left on the long drive to town; I followed later with Grandma who was visiting for the birth.

Then: long hours in the motel room with Grandma, longer than anyone expected. Dad darkening the door and leaving. Dad and Grandma's conversation that I couldn't understand, except the sense of something tense. I tried to settle myself on the hotel bedspread, tried to animate it somehow, turn it into hills and valleys.

My brother was positioned breach and a pound bigger than I'd been. Her doctor was not an obstetrician. There *was* no obstetrician, and the baby's shoulder was stuck. After a long time pushing, she'd had enough of a doctor who didn't listen and seemed to understand little about the process of childbirth: She yelled at him to get out so that she could deliver the baby without him. Andy was born after an extended, damaging labor.

* * *

Grandma stayed with us for the first week. And then Mom was on her own in a cabin that still lacked indoor plumbing, cooking dinners from scratch and, out of her own sense of inherited domestic mission, making sure we had freshly baked cookies on hand. She tried to stitch the distance between herself and Grandma in a letter written when my brother was about two weeks old.

She catalogues the ways she is picking up the mantle of domesticity now without help. Both babies are asleep; she knows she should be napping too, but she's baking cookies so that we'll stop asking when Grandma's coming back. She admits to being tired, but *what I am here to do*, she writes, *is give three other people all the love and good food they need.* And she describes exactly the strange calculations that took, living the way we did.

> *It's pouring down rain. It had been beautiful for almost a week but yesterday I simply had to go to Baker for food. I packed my two littles in the car and headed to town in 85 degree weather. By the time I got there (stopping twice to nurse Andrew) and put the laundry on and done my downtown shopping with Andrew in the belly pack and Jessie having her usual ½ hour conversations with the store keepers I was so hot and tired I threw the wet laundry in the back and bought a clothesline and some pins at Safeway. Safeway was fun. I put Andrew in his carrier in the basket and Jessie wanted to sit in the seat of the shopping basket too so she could teach Andy to shop. By the time we got threw the fresh fruits and veggies the cart was absolutely full—you could just see Andrew's face between the corn and tomatoes. Luckily (sort of) Andrew woke up and Jessie wanted down so I could fit the rest of the groceries in the cart. But since Andy was now starving I had to race around at full speed with this crying baby and Jessie kept getting lost down the wrong aisle and would be calling desperately from the peas while I was*

in soups wondering if I could inconspicuously nurse Andrew while pushing my cart thru Safeway. By the time we escaped from Safeway and fed Andrew and got headed home I could see the clouds gathering. It has now been raining for a solid twenty-four hours since I hung the clothesline and the bottom of the jeans are about a ½ inch above the ground.

I remember the grocery store episode she describes: finding myself alone in the aisles, mother moving on much more quickly than I was used to, not knowing where my body belonged around the nucleus of her and the baby. The terrifying feeling of being in a big place with no bodies to attach to.

My mother has told me a story of an event I remember. I didn't fully understand until I had two children of my own. The summer when Andy was a baby, less than six months old, the plumbing wasn't in the cabins yet, but in the assay lab, a shower had been installed, and she was desperate for it. She wheeled Andy and me in the old-fashioned buggy from Leadville down from the cabin and parked us outside. She told me she'd just be a minute, and to stay there. She's told me that there were various men around. I remember the sense of adult presence if not any specific people. I remember being in the buggy and wanting to get out. Normally that wasn't something I was supposed to do by myself, and I wasn't sure if I *could* do it. But something about just sitting there with my sleeping brother seemed impossible. I remember stretching one leg down out of the buggy. I remember the sick feeling when the whole thing toppled over.

Andy's face was cut on the crushed rock that paved the camp, a horizontal tear just below his eye. Mom rushed out of the lab in a towel. He needed stitches, which involved an hour-long trip over the bumpy road to town with a bleeding infant, and of course, me. He still has a faint scar on the spot.

When she came out of the shower, the men were standing around, inattentive to the two screaming kids, one an infant. They were disregarding us. Because they were men and we were children, or maybe because my mother was the boss's strange wife who lived in camp even when she ostensibly could have lived elsewhere, and we were the boss's children. Either way, she didn't experience even the most basic sense of solidarity.

In that summer's letters, Mom says I, as a toddler, was spending *too much time alone.* With my father in *one of his work-a-ton periods, I feel badly,* she writes, *when I realize she has spent most of the day by herself and I have had to say no too many times when she has asked for a few minutes of my attention.*

Several months after Andy was born, the plumbing finally went in. By then, my parents were already considering their next move.

Winter came again, and more heavy snow. My mother wrote, *I keep expecting it to stop snowing and it just won't.* We were banked in for weeks, and she worried about running out of food before the county could get the main roads plowed.

The blueness of the sky when the melt began seemed to have a near-human agency, seemed a furiously energetic depthless being. When the melt finally came, I wanted it to be summer so badly I convinced myself it was and tried to "go swimming" in an icy mud puddle. I was chasing a sensation that came on inexplicably, when the world—which is to say, what mattered—seemed extremely close.

One afternoon I was standing on the steps of the cabin, silent and breathing, having just finished some proclamation, some gesture of ferocity, some piece of dialogue, some intonation I had no chance to use with anyone and had to practice with the dog, the dog who lacked the human features I desired in a companion, whose hair was short and could not be braided or curled, or if not with the dog, with my worn cadre of small, stuffed avatars: two birds, one rabbit, one bear.

The steps descended from the cabin's front porch. My mother was behind me, with paper and pen.

In front of me, a field of crushed rock, a stand of fir trees large, ragged, and strange in the foreground, a sky growing serious, dark clouds in one direction and a strange golden light in another, the sunlight filtered through a narrow opening in the clouds, all the colors warming and intensifying: the conifers goldening, the underside of clouds bluing, our little yellow truck the color of a pale yellow apple.

I fell from my waking dream into the right-now, struck by the sense that this, *right there* where we were could in fact be a setting, a setting in which we might be the main characters, a sense that the story we were living right-there-on-the-mountain meant as much as a trip to town, as much as anything else we might read or access or acquire.

A bald eagle floated down slowly on enormous wind-tipped wings, so slowly that when my mother said *look*, I could see the bird, the feathers, the controlled descent to the tree. And my sense of momentousness was so great that I told my mother, *I want to write a poem* and I named each thing in the scene, each event—the sky coming in, the colors, the eagle—for her to write down.

As another spring brought about a general emergence of people from their cabins and trailers, my mother sensed trouble with the townsfolk in Granite—a kind of trouble she wasn't used to.

> *We went to the Granite bimonthly blowout last weekend. All local bluegrass pickers, partiers and prancers out in the field by the store. There were probably forty people who drank five kegs of beer and assorted other beverages and I decided this community is too small for such incredible goings on or everyone will shoot each other before winter. However, Marshall Bud was there with*

his six shooter looking worn and well used on his hip. All stayed in hand.

Marshall Bud was not in fact official Granite police, though he may have been a retired lawman of some kind. He was the self-appointed mayor of Granite and made it his business to be armed and helpful. He liked to show visitors the ghost town and tried to tamp down the kind of behavior that might scare them off. From Vance Bennett's account, he took care of people, too, pulling the cars of drunken folks out of the ditch, trying to find places to sleep for people who couldn't drive out. He lived alone in an A-frame cabin a little way toward Sumpter and mined on his own when he wasn't marshaling.

Mom felt the threat of impending violence without being able to say exactly why. Now she says that "some of the people seemed a little off."

On the eve of 1981, fifteen months after we'd arrived in Granite, Mom put Andy and me to bed and stayed up late with Dad to plan the future. *We made quite a list of resolutions, too; but the biggy, getting into a home and working situation we all like is still very hard to see taking shape.* Home—as in the sense of balance and steadiness—had become a perpetual, unresolvable problem.

She mentioned going back to school; Dad again thought out loud about other things to do besides mining. In the end, we ended up somewhere even more remote.

The Company had another mine in southern Idaho near Jordan Valley, and that spring and summer Dad was back and forth between the two places with some of the Granite miners. We all left Granite in the summer of 1981, when I was three and a half, for a camp on South Mountain.

Dad went ahead of us. In my mom's last letter from Granite, to Bev Walker, she was planning a trip to Portland for her sister's wedding before moving to a camp not close to anywhere, a camp that offered nothing in particular—no house, no cabin. Still, she found ways to look forward to it.

George has been gone, except for brief visits, for the last two weeks. Today all of the miners and their families are moving down to Jordan Valley too so the mine will be in full swing by Friday. George said the camp is knee deep in mud and cold but the houses are all in, gas stoves and lighting, a shower house with two johns, a wringer washer, and three refrigerators that will serve the whole camp. [. . .] The next six weeks will be a hustle and I'm sure that by the time we all troop back to Jordan Valley after the wedding the little tent in the woods will be a welcome sight—it will be much more fun to explore around our summer home. I hope that some combination of people dear to our hearts will be able to visit this summer. The Malheur bird refuge is close and one of the best in Oregon complete with hot springs and the delightful French Glen Hotel. (Close of course is about 70 miles, nothing is really close to South Mtn.) On the Catlin field trip last spring the group sighted 145 different kinds of birds there. Speaking of birds we have had, are you ready, Sandhill Cranes! around here all spring. Such a marvelous prehistoric looking bird. The blue Heron has been here too and his elegance made the cranes look even larger and more like a feathered teradactyl.

Give my love to all over there and we will see you in a matter of weeks. Come as early as you can. You can help me pick out a dress. I haven't bought a dress up dress since I was in high school! I don't know what to look for anymore.

The "houses" she mentions were mine-buildings, not for living in. She writes, "our summer home," which meant a tent made of canvas and two-by-fours. She and Dad had been talking for years about ways to step out of this way of living, but she kept stepping further in.

The camp she called Jordan Valley in this letter was up a long dirt road at the top of South Mountain, Idaho, in the Owyhee Mountains. In the 1870s, the settlement at South Mountain was called Bullion City, then South Mountain City. Now those places merit a just a slim paragraph on Idaho history websites. On Google Maps, it's beige and green in every direction. Zoom out and zoom out, and finally a few placenames come into frame: Flint, Cliffs, Triangle, Dickshooter. Click on any one of them, and you'll see nothing but sage.

"Town" from there was Jordan Valley, current population 130, where there was a Basque butcher who had fresh meat hanging and would slice off exactly what you asked for. We lived in a canvas tent, then a trailer, and kept the tent for a place we could escape to in the heat. Andy, exploring by then, contracted giardiasis from sucking on rocks in a creek that turned out not to be pristine. To have it diagnosed, my mother had to drive stool samples to the distant hospital, an all-day drive round trip, and return a week later for results. To treat it, she had to get him to eat a bitter pill multiple times a day. It took two adults to get it down him.

In hindsight, my mother says, "God, what a miserable camp that was."

I remember baked dirt, miles of sage and skunk cabbage, sun-hardened cow pies. Men gathering around the trailer. Me, not sure of where or how to be.

I remember shortages of particular food items: asking for juice and not getting it, and my mother being suddenly and surprisingly angry at me for wanting it. She was past her limit there, whether she knew it or not.

From the South Mountain camp, she sent often-unanswered pleas to her mother, hoping for vacation, hoping for a visit, asking for a letter.

The previous mental escapes and romantic self-conceptions of the previous letters were not available here. The camp was extra temporary in that it was owned by the same employer Dad had worked for in Granite, and we could, in theory, be sent back at any time. There was no possible cabin, no fantasies of homemaking in that camp, no artful housekeeping. There are no mentions of breathtaking landscape, no schemes, no plots.

She was cooking for our family but also whoever happened to be in camp. It was hot and buggy, and we all struggled to stay well. Trips to town were harder to make, and she ran out of diapers. I outgrew my shoes and couldn't go far in the sage.

My grandmother wasn't reliably writing back, and my mother started begging for some cheering up, some good news, some wisdom. She imagined Grandma having some kind of difficulty that might have prevented her from writing. In a letter, my mother—needing help herself—says she wishes she could be more of a help.

The letters show a move that I am by now expecting: pronounced difficulty paired with resolutions and a compulsive minimization of the difficulty: *We have only had one night of storms, but we found out that our tent is not waterproof, not even a little bit waterproof. But since then we have had no clouds at all.*

In the South Mountain camp, the men's movements again seemed to make the world go round; the clock depended on whether Dad worked nights or days. When he was on days, he got to follow the blaze of morning out into a purposeful activity. I remember the trailer door swinging shut behind him. I too wished I could exit.

Men sat around outside the trailer. One of them whose name I didn't know was sitting on the metal-grated steps while my mom was cooking and I was inside without clothes on, and I had to pee, and I thought I could pee standing up (as was my habit) right over his shoulder, and so I tried that. But I miscalculated and splattered warm liquid on his brown plaid work shirt. Mom was angry. I was indignant. No one ever taught me how to ask a man to step aside.

My work all day was imagining. I walked a little way away in my too-small sandals and steered my Fisher Price dune buggy through the sage. My mind suddenly telescoped down into the field of my imagination, turning a sage tuft into a tree, the buggy's path into a road out.

At the end of a long summer season in South Mountain, we moved to Town: Baker City, Oregon, the county seat of Baker County, population ten thousand.

Dad was still working between Granite and South Mountain, but the camp in South Mountain wasn't fit for winter, and in Granite a paranoid mother-daughter pair under the influence of a drifter from the East Coast had shot Marshall Bud, their next-door neighbor of decades, over a series of imagined slights. In a rare concession to safety, my parents didn't see going back to that cabin as an option.

Baker was a place I knew from the long-awaited trips down from Granite. Baker meant a playground, restaurants, running water, sidewalks, a pizzeria, an ice cream shop—and people, so many other people.

My parents bought an old Victorian house on a big corner lot. The furnace was dangerous, and the price was right. They brought me with them on the day it closed. We ate at a pizzeria and I held their hands as they whisked me onto curbs in the exhilarating street-lit dark.

There are no newsy/anguished letters from the eighteen months in Baker. For the first time, there was a phone.

It was the first real house I'd ever lived in. I could drift from room to room, looking up at the locust trees planted outside. I could venture into the yard. It was twelve hundred square feet at most, but it felt enormous. The house was a place made specifically for people like me. It was not a jobsite. I remember the door handles, the curved molding, the high windows, the light fixtures. Instead of imagining a real life far away, I woke up right in it.

Dad was home only on weekends, but my parents devoted their considerable DIY industry to the place. They replaced the house's old systems, painted it classic white, and had friends—suddenly so many, suddenly so near—over to barbecue. I went to preschool three days a week and my mom combed yard sales, acquiring what seemed like an incredible abundance of objects. We got an old black-and-white TV that barely worked but worked enough to be magical. Mom had a sewing machine and a place for it, and she made me a pink-ribboned Easter dress. We raked maple leaves into piles behind the picket fence. We trick-or-treated in homemade costumes.

I turned five there, but not six. There's a picture of me buried in carrots and potatoes from the huge garden they planted. We stayed for a single harvest.

• • •

As Kevin and I grew used to the new house, the anticipation and excitement of transition faded and the days blurred together. How often did I wish there were a way to still our world just long enough that I could see its shape.

Clementine's hair in the sun, curling gently at the ends as she ran across a concrete footbridge behind the house.

The optimism and certainty in her postures that I didn't know how to capture in words or images.

The seriousness of all her play, as if she had no doubt about whether she belonged in the world and whether she'd be powerful in it.

Clementine in the car on the way home from the nanny share on a darkening day, sighing. "Look at this sky! It is turning a beautiful blue for us."

And Kevin playing piano, first gently with infant Clementine laid across his legs, then sitting on his lap, and then beside him. A new lullaby or lyric emerging from his guitar or mandolin. The exquisite putting-to-bed melodies that would be played once and never again. The everchanging song of him, none of it written down. I'd try to record it, walk out to get closer and interrupt it, listen to it later and find the recording warm and beautiful.

In any particular day, the hyperverbal child said too many words. Too many things happened. This sweetness could only pass me by. The elements of context scattered before I could understand.

* * *

I became pregnant with Paul just after we moved into the new house.

One day after work, I walked up the wide yard with Clementine's hand in mine after taking the call where a genetic counselor read me the results of the chromosome testing. I called Kevin to let him know we'd have a boy. My children would be in the same configuration as my brother and I were: older girl, younger boy, about three years apart.

I walked in the door and collapsed, lolling in screen time and letting Clementine be near as she let loose a stream of speech that I let wash over me.

Each day, the job edged past every boundary I tried to draw, and my body burgeoned past the well-tended outlines I'd fretted over my whole life. My growing torso ate every piece of energy I had, hated food and craved it, just wanted to sleep and couldn't really sleep. I got home well after dark and walked this body—which suddenly recoiled from exercise even if my mood desperately needed it—through the dark, twinkling neighborhood.

On Halloween that year, I ducked out of work "early"—actually a regular leaving-work time in the late afternoon—and managed a last-minute trip to Target. I tossed through the nearly bare shelves, thinking of my mother, who would have started weeks before on a handsewn creation or crafted some ingenious thing out of old packaging and electrical tape.

At home I put on the adult bunny ears I'd just bought and convinced two-year-old Clementine, who was locked in the forever-now and had no idea what I was talking about, to put on a pair of child-size bunny ears and try this thing called trick-or-treating. We snapped a selfie in which I tried to reach down and smile as the nonpregnant person within.

We made it to four houses. At the first, she rushed into the living room, and the homeowners laughed with me. At the last, up a rise of jagged steps with just a single porch light on, a small gray woman opened the door. Clementine dazzled with her bunny-nose twitches, her songlike voice, her precocious pronouncements, her fierce self-proclaiming.

"This is your daughter?"

"Yes, and a son on the way," I said, patting my stomach.

She straightened up and looked me in the eye: "You are so, so lucky."

We thanked her and left and walked home.

We carry the idea of us around with us. I was born to a people living far out. The idea of me formed far up that long road. She was in me still, an inconsequential person alone in the long-too-hard, a long-too-hard that I somehow kept finding, not knowing how to see what its opposite might look like, not knowing the magic words or deeds that would release me from a state of endurance.

She was right, this woman who must have been lonely, a person I might look in on if I was the kind of person I wanted to be. This abundance in my life, the profound good fortune of being able to nurture, even imperfectly, this bright, strange child. I wanted to be able to feel it.

Two months before Paul was born, I saw a Groupon for a sensory deprivation tank. I longed to float and thought that if, in the absence of stimulus, I could perceive something ambient noise normally drowned out, so much the better. Uncomfortable in the outside world, I could come home to my own thoughts.

In the converted living room of an old house, the hippie told me what to do and then showed me to a dim room upstairs where the tank sat like a fat, plastic clam. I showered and then climbed inside, naked and gigantic. I pulled the shell down over the last small gaps of March daylight and pushed a button to turn off the internal lighting. The tank was narrower and shallower than I expected. Still, supported by a rich, salty fluid, I could float without touching anything. The sameness of temperature smudged the difference between the water and my skin.

Some floaters report visions and hallucinations rising up to fill the absence of sensation. For me, the movement of the baby filled the empty space. In his motions, I felt an insistence, the pressure of a separate being. *He will be himself, his own person,* I thought, *with his own plans and fierce momentum. He* is, *and he will change things.*

I don't know how much time passed before the tank woke me from a sleep-not-sleep with a bloom of light and soft music. I showered, dressed, and stumbled downstairs. I stepped out into spring rain and street sounds. As real life collected around me, the memory of floating faded, except for my impression of the baby's strength and energy.

Paul was born fast and without drugs in a rushing wave of fluid. In a way that Clementine had not, he insisted on his attachment to my body, day and night, minute by minute. Like me, Clementine came into the world looking around. Like my brother, Paul arrived with his eyes shut, certain of the dark warmth he needed.

After Clementine's birth, my cheeks pinked. We hiked days after coming home from the hospital. In the pictures taken in the days after Paul was born, I appear to have lost all the blood in my split, bruised, and stitched-up body. The muscles responsible for holding my hips together and swinging my legs no longer worked. I strapped him to my chest and sat inside the house in the warming weather and didn't make as much milk as he wanted to drink and soothed him through his stomach pains. Clementine adored him and grew inflexible in her preferences. She couldn't understand why I was unable to pick her up.

The summer he was born, it thrilled me to make short drives with both in the back seat, foreshadowing road trips and fun errands. But Paul plunged us into total parenthood. In our first weeks as parents of two, Kevin and I crossed paths in front of the refrigerator. Our eyes met briefly before we diverged to meet a child's urgent need. There were no more discussions about the division of household chores, no discussions of anything, really. Both of us worked at work or kids or housekeeping without a break.

When Paul was three weeks old, I got a call from a press I loved. The book of poems I'd been submitting off and on in various versions for years, sustained in the intermittent periods when I was able to think of myself as a writer, had won a chapbook prize. To take the call, I shifted Paul to my best friend Jenine who was visiting from Seattle. He cried like a pissed off pterodactyl through the call, red and writhing.

I thought of chapbooks as signifying promise, and having one was an affirmation. But everything I'd learned in school about writing, about being a writer, was study and control, self-disappearance in service of craft. Now I was a beast whose body could make a tiny human's food, an animal who could survive with little sleep, a worn-down fish programmed to keep going. I had no idea what, in this condition, *promise* might mean.

Summer: during the day I'd bring Paul outside and lay him in his cradle in the shade of a patio umbrella. Tall cedars ringed the small yard at the edge of the ravine, still overgrown by invasive holly trees and English ivy that choked out the native mahonia and sword fern.

I would squint to read on my phone. In the absence of work to occupy my mind, the horrors brought by the newsfeed hit harder, stayed longer, and implicated us more.

That summer, police choked a Black father to death for selling cigarettes and killed a Black man about to start college because he was walking in the middle of the road. Nothing new about it except video, but video had consequences, making visible not just what had happened, but the way people reacted to what had clearly, certainly happened. Reels of beatings and deaths filled my feeds unbidden, and my baby learned to smile. As I nursed and read articles, nursed and signed petitions, nursed and donated whatever I could, the comments filled with vocal and numerous strangers and acquaintances responding to state violence against unarmed human beings by denying what was in front of their eyes, making excuses for the inexcusable.

I remembered the new rage I'd encountered as a sometimes slow-walking pregnant woman shopping in the suburbs—the suddenly overt rage of white men slightly inconvenienced by someone they found worthy of contempt. And I knew that what I saw and felt was nothing, relatively.

The fight was here, and a bigger one was coming, the old cycle of hatred amplifying, and I was attached to a delicate animal who nursed every hour, whose crying spells I couldn't always figure out, whose fontanel pulsed delicately. We had a little money and a few words to throw toward the right side. It didn't seem like much.

Again my postpartum imagination was sharp, hard, and inescapable, like an ancestral artifact from times when many, many infants died. When Paul cried, he cried like it was an emergency. One Saturday I begged Kevin to take him—we both knew how he might scream peeled from my body—so that I could go for a walk, and I tried with each step to stay present in the here-now of that moment, placing each foot gently, feeling each sun patch and leaf shadow before it passed away.

How can I care for him, for anyone, if I can't act beyond this enclosure.

And then, there was nowhere else to go except back to my particular circle of trees, the sprawling house and large lot, disconnected. At least not that day. I told myself there would be others.

At the end of the summer, just after my tenure letter arrived, I went back to work again, full-time, and as a department chair, a scheduler and evaluator of other teachers. I knew I would need something to wear, so I bought two pairs of pants online. I tried them on and they almost fit.

Tenure changed nothing: the job reached into home and home reached into the job and a bloodless exhaustion underlaid every scene, except one. When I was in a room with actual students, the demands of presence awakened some person inside me who could listen and think and speak to the voices in front of me, a person who reflexively cared about the bodies in the room and could sometimes find ways to express that care in action. In the immediacy of teaching, and only there, I was fully paying attention.

I forgot to renew my driver's license for a year. The letters from daycare and preschool about the children's basic seasonal needs—*please send a warm jacket, hat, and gloves*—took me by surprise and plunged me into panic. When would I have time to find these things? Where could I buy children's gloves? I ordered them online and they did not quite fit. A lost water bottle could set my chest wall hardening to tin, start my heartbeat pinging painfully against it.

Out of defiance and stubbornness, I kept writing, no longer with my mentors in my mind, no longer with anyone in mind, but as a beastly reflex. I still wrote in the four o'clock hour, but this was no longer before the rest of the day started. From another room, Paul could smell that I was awake and paying attention to something besides him, and he often woke up crying. He wanted me unequivocally. His father's body didn't comfort him. I wrote lying down on the couch, looking at the screen out of one eye while Paul slept across half my face. If I didn't finish grading or planning the day before, which was often, I used the four o'clock hour for grading or planning. If the children were wakeful at night—if say, they had colds—and I'd just gone to sleep at 2:30, I slept through the alarms.

But I kept the alarms on. I sailed through the days on caffeine, momentum, and the sense of something to prove, forgetting where my keys were, when the children's dentist appointments were, when the school pageant was, and feeling every fine fluctuation in my students' and coworkers' anxiety, anger, and joy. In an interview with a potential new hire, hot tears covered my face when I talked about how much I loved my job. What I would have said, if I'd had the words, is that I felt it, hard.

In the fall, when Paul was eighteen months old and Clementine was four and a half, when I was not pregnant, not moving, not starting a new job—in the season that I imagined would be my return to normal—Kevin woke one morning drenched in sweat with intense abdominal pain, white-faced.

We thought appendicitis, but the pain was on the wrong side. I tried to talk him into letting me pack up the kids and take him to the hospital, but he drove himself in the still-dark, and I called his mom and asked her to meet him there. She called me later from the ER to update me on his progress. He was in an exam room, but they knew nothing. I could hear him screaming.

I made the kids' breakfast and got them to preschool and daycare, late. I believed that they could know everything and not feel it too hard as long as I kept the basic pattern of their days more or less the same. I believed that if I kept mealtimes and school times and bedtimes and bath times steady, I could keep their world together.

I got to work late and tried to focus, sitting at my desk and making list after list of what I had to do before class. I waited for news and thought about when I could possibly get to the hospital. *Can I grade these papers at the hospital? Do they have Wi-Fi?* I had no language to stop the week's multiple deadlines, or even create enough time to figure out a plan.

Kevin's mom sent updates by text. *They're doing tests. He has been admitted. His pain is not under control. His pain is a little better. He's vomiting. They can't stop the vomiting. When can you come?* I'd been through hospitalizations with Kevin before, but this seemed different: more severe and less comprehensible. Abdominal pain was not on the long list of autoimmune symptoms we knew to expect.

The academic quarter, at least at community college, doesn't stop for illness, emergency, or bereavement. It doesn't stop for court dates, incarceration, or loss of housing. It doesn't stop for trips, conferences, custody hearings, or mental illness. It doesn't stop for broke-down cars, childcare emergencies, or lack of funds for a bus ticket. For the students' problems, which are frequent, I do what I can to make little pockets of amnesty for them to take care of themselves without lapsing too far behind. My absence, though, affects dozens of people and throws off the schedule, carefully arranged to give students just enough time to learn the things they're supposed to learn. For me, there was little amnesty, not much backup—not any way to know what help might look like, exactly.

Class over, I canceled office hours, picked up the kids, and headed for the hospital, following a string of text messages through which a story emerged.

In the previous days, Kevin, following one doctor's orders and without saying anything about it, had weaned himself slowly off the dose of corticosteroid he'd been taking for years to suppress his overactive immune system. He'd started taking it years before, during an episode of hemolytic anemia (in which his immune system destroys his red blood cells faster than his body can make them). Better drugs were supposed to replace the prednisone; the better drugs didn't work and so he stayed on what was supposed to be a stopgap treatment.

We'd heard it before as a casual aside from specialists and generalists: *Oh, and you should probably get off the steroid*, which weakens bones, joints, and connective tissue and can hide the signs of a smoldering infection until it grows severe. Every time he tried to taper off, he'd get sick and go back on it. This time, though, a new rheumatologist told him to try again, and he made it all the way to zero.

The day after his last dose, he woke up with the howling pain in his gut. Throughout our ordeal, no doctor would speculate about a cause-effect relationship between the medication and the infection.

Tests showed a large patch of inflammation, red on the screen, which looked to be diverticulitis, a common enough intestinal infection that he was young to get. Kevin's lupus, though, had led to massive internal infections before without apparent cause, including a sudden case of sepsis, and his mom and I wondered aloud if lupus could be involved.

I made it to the hospital, the same one where the kids were born, in the afternoon. Kevin was heavily drugged. His eyes fluttered open for a minute. He grinned at me and passed back out.

On that first day, or maybe on the second, a surgical intern visited us. She was very short and very young, and at first I thought she might be a kid-genius Doogie Howser type, but then I realized she was also heavily pregnant. It seemed she had not yet been taught what not to say, because she proceeded to articulate, rather than discount, what we'd all been thinking. A question was coalescing around whether or not to surgically remove a diseased section of his colon. Entirely without expression, she said that if this inflammation was actually caused by an underlying autoimmune disease process and they removed a section of colon, the inflammation could return and they'd have fewer options.

"If we just start cutting, you could wind up with nothing left." She whisked out of the room.

Kevin has a long, purple scar down the center of his chest, a vertical line widened by horizontal stretching, where his thymus was removed at age fourteen, weeks after his myasthenia gravis was discovered—the autoimmune disease that causes his skeletal muscles to weaken intermittently and fatigue easily. He has scars on both ears after three surgeries to repair eardrums blown out by an unexplained series of ear infections. They ruptured again, and he opted not to repair them. He's had pleurisy and pericarditis, both from lupus, the secondary diagnosis that causes an amorphous set of symptoms that are mostly chronic and occasionally acute. He has shut his eyes as doctors pushed long needles through his back to extract a piece of liver or draw out a stream of bone marrow. He has many times imagined his body being cut away until there's nothing left, imagined the flares of inflammation moving into areas more important than an eardrum. A death from lupus happens one strange inflammation at a time.

On Kevin's birthday in October we walked about a mile with the kids to a park with swings. He'd been out of the hospital for two weeks. They'd pumped him full of IV antibiotics, monitored him for several days, and sent him home with more antibiotics, which he'd just finished. Leaves made shapes along the park's curved pathways. Beams of sunlight glanced off the iPhone photos I took of Kevin pushing the kids in their swings. His arms spanned the distance between the two hard rubber seats. His large hands tapped each smiling child higher and higher and higher.

I thought we might be back to regular life, which would be more than enough, even without a solid sense of what had happened. I would finally get some sleep. I could finally catch up on the piles of papers yet to be graded, the emails waiting to be sent, the classes to be prepared. I thought we might have friends over for dinner. Maybe we could plan a day trip to the coast.

The kids needed lunch and Kevin walked off to get quesadillas from a place across the street. The kids and I waited and waited. Kevin's face was sickly white when he finally stumbled toward us, sweat beading on his face. "It's back," he said, "I have to go in."

Over the next three months, Kevin would be hospitalized six times, sometimes for days, often for weeks. He was in the hospital more often than he was out of it.

At first, the doctors were trying to heal the infection well enough to test for Crohn's or other autoimmune gut problems. Each time he was hospitalized, he'd be released with antibiotics to continue at home. As soon as the course was complete, the infection came back, stronger, it seemed, each time, before any tests could be done. The healing never happened.

We always thought he was about to get better. We could never see where we were.

One of the recurrences was worse than the others. He was admitted to the cancer ward where there were extra beds. The ward was new, the room ominously spacious. The crucifix—it was a Catholic hospital—the smoothest kind of modern, conceptual to the degree that a body on a cross looked almost like an orchid.

I arrived after Kevin's mom said, "You better come." In this room, Kevin's vomiting wouldn't stop, fever wouldn't come down. As always, we waited for doctors. The stack of papers in my bag went ungraded, again, which just meant they would be graded during hours when I ought to be asleep. Being caught up with everything: I chased this condition. If I was caught up with everything, I thought, I could rest. I couldn't let go of this idea, this condition that never arrived.

Kevin's mom and I made plans about who would pick up kids. The nurse looked at me and said, "I didn't know you had kids." She was blinking away tears.

That night I got the kids and dropped them off at Kevin's mom's to stay the night. The routine I'd tried so hard to keep intact was fraying. I stopped into a store on the way home, not the usual one. It was an old grocery whose parent company had just gone bankrupt. It was about to close, permanently, but it was not closed yet. The signs were all down, the inventory stacked in odd places.

Still, I thought, *I can find what I need for the kids' lunches, a little coffee for the morning*. A small jug of apple cider attracted me. It would go great with bourbon, warm.

I passed others in the aisles, each in their private-public selves. The other shoppers were still a part of regular life, where nothing is at stake, where love continues, and I was in my own dimension where a loss I hadn't dared to imagine seemed—suddenly—possible. *This is what it's like to be a person who is losing someone.* And then I thought I ought not to assume. Maybe some of my fellow shoppers were there with me in the numb adrenaline nightmare of Loss Land. Maybe the idea of regular life is the exception I'd been lucky to inhabit all these years. Maybe we were sharing this same story after all, me and the people, mostly middle-aged and old, wandering up and down the aisles—the story of loving and losing each other. Maybe I was just late to it.

Kevin finally came home a week later, after yet another course of antibiotics, depleted from pain and hospitalization. I crawled into bed next to him. The fact of his mass, his being, his existence overwhelmed me, and I burst into tears, and he asked why. "You're just so much. And you're here."

It was a long time before I recognized how badly I needed her, but when I did, my mother came to stay for weeks at a time. She washed and dried and folded laundry, made dinner, retrieved the kids from their preschools. And between her and me and Kevin's mom, the kids' routine came back into focus, with the addition of a prebedtime video call with Daddy during the periods when he was in the hospital.

I canceled only two classes. All the papers got graded. We didn't sleep. The baby got sick, and I got sick, and I was up all night with him, his heavy bald head pressed firmly into my neck, and then I was up late grading, and then I was up early grading. I leaned heavily on a text message thread with my friends from grad school, also poet-moms, who knew and loved Kevin, too. I sent out updates and they sent emojis and jagged, hilarious encouragement. The thread was a place to put my words when I felt like I couldn't stand the uncertainty and exhaustion—when I needed an audience for it. Writing to them turned me, briefly, into the protagonist of what I could briefly imagine as a story.

But like Mom in her scenes of isolated desperation, I lacked the understanding, or the option, or the words, to create the conditions for rescuing myself for more than a few minutes at a time. No matter what, I kept the house, the schedule of meals, the peaceful bedtimes, the stories and scheduled treats. I maintained these rituals of the house as refuge and rest for others when I myself needed refuge and rest.

My cough blossomed into chest pain. My daily resolutions to accomplish tasks revved my heart. A chill in my center craved warm food cooked by someone else.

One day, when I had a minute at the dining table to read students' papers, the last part of each word vanished behind a spot of black in my vision. Internal pressure had made a hole in what I could see.

Somehow, on one of these days when Kevin was in the hospital, Mom and I managed to go for a walk together before one of us had to go get the kids. We swung our long legs up the wet street. She asked how I was doing, and I said, "I feel like I'm supposed to be in three places at once. At home. At the hospital. At work."

Kevin's illness, which left me as the sole capable parent and added an extra location to the already strenuous work-home-double-preschool situation, crystallized the preexisting general feeling of being eaten up and being nowhere at all.

It hurt to breathe. Months later, I'd see the shadow of pneumonia on a chest X-ray.

"Well," she said through the dripping fog, "you can only really *be* in one place at a time, right? That's one thing to know."

I knew what she was saying: to drop down into the moment and stay there, not tracking times or tasks, not trying to mentally chop up and squeeze a huge, tangled mass of labor into the insufficient time I had for it. It was familiar advice. I could see myself in a clearing, in a strange alpine grass, a place that only I could make. I longed to be so present. I couldn't imagine what would happen if I let myself be so, if I gave up the mental landscape of deadlines, of forecasting and adjusting and trying to make everything fit. I feared giving up that effort. The effort was one of the few ways—amid the cacophony of needs—that I knew how to feel my own existence.

In one of Kevin's convalescent periods, I arrived at the hospital in the afternoon, in the middle of a dark, pounding rain. I pulled into the garage and watched drops falling off the rhododendrons planted just beyond the structure, as if the windshield were my personal theater on the strange colors made by a heavier sky. I walked into thc hospital and got on the elevator with two other people. One was a small woman, a well-dressed visitor, older than me. The other was a young hospital employee. The doors closed. I look down and exhaled, cold again, always cold.

"I. Love. The rain," the woman said slowly, as if talking to herself. "I'm from California. I love the way it pushes you inside, under the blanket, with a book."

"I love it, too," I said, nodding without looking her way.

"Lived here all my life," said the boy. "And I still love it."

The bell dinged, the doors opened, and the woman and I both exited on the eighth floor, going separate ways.

When I reached Kevin's room, I peeled away my jacket and shoes and crawled in beside him, pulling the thin, bleached blanket over our legs. The crucifix on the eighth floor was the older style, that molded kind that gives each of Jesus's glossy waves curve and definition. The loudspeaker intoned a daily prayer and I remembered that down the hall there was a chapel, and in the lobby there were pictures of nuns and a history of the Sisters' medical work in the Northwest. The religious veneer over events suggested another dimension in which we could pray, in which we'd have the solace of an idea, a concept above this tenuous mess of empty pudding cups and IV bags and hospital socks, something beyond the doctors and their limited insight on the doomed and persistently mysterious human body.

I rested on Kevin's ill shoulder, snug in the ill arm he held tight around me. He tried to make me laugh. He worried about my cough and admonished me to see a doctor. "Don't mess around, Johnson."

For an hour or two, we talked as we hadn't since Paul was born, like we were on a date, without work to complain about, without plans to sort out, with only ourselves and each other, our congruence and our pact, to lean on.

Throughout Kevin's illness, the alternation of crisis and recovery, wave after wave, localized me, then tore me apart.

One night, I wound the car home through thick fog, listening to an interview with Cary Fukunaga on the crafting of the *True Detective* TV series, whose moodiness and existential questions absorbed me during my maternity leave with Paul, and it seemed that I was in a noir film, the city awash in rain, the hidden forces of Kevin's biology shifting our fates in an unseen realm. This feeling of having stepped into a story stayed with me for merciful days, and I became again an imagining being instead of an ever-failing mom-employee. I could imagine myself-the-ever-failing as a protagonist of something. When things got too hard, my mind knew how to rescue me by making the world into something else.

Finally, in early December, Kevin was having surgery and it was all about to be over. Unable to get him well enough for the tests that would rule out Crohn's or another autoimmune gut illness, the surgeon decided to operate on him anyway.

I waited all day with his mom in the waiting room until we heard: it had gone well, there would be no colostomy bag, it indeed looked like a severe enough case of diverticulitis to cause all the problems it did without an underlying autoimmune process.

They cut out a piece of him, he would heal this time for sure, and it was not supposed to come back. The quarter was about to end. We had just enough time to get ready for Christmas, a key element in the annual home routine that I understood it was my responsibility to maintain. I thought we'd made it.

The next day in Kevin's hospital room, an occupational therapist who looked like one of the male members of Peter, Paul and Mary breezed in to talk about a recovery plan. He was full of rehearsed jokes and metaphors. He thought of himself as *a fun guy,* I could tell.

Kevin wouldn't be able to lift anything heavier than a gallon of milk for six months. The therapist talked to him about navigating his job, about when he could drive, about how to get out of cars. Then he asked if there was anything else. My chest hurt; my heart raced.

"We have two small children," I said, "eighteen months and four. Both of them weigh more than a gallon of milk." *I think I have pneumonia,* I considered reporting. And a full-time job. And I hadn't slept nearly enough in months.

"Oh!" said the man, turning toward Kevin. "Are there any activities you enjoy doing with the children?"

It seemed he was unable to imagine a household in which parenting is a shared responsibility, unable to imagine that our house only functions because Kevin, when he's well enough and when he's not quite well enough, does laundry and dishes and puts both kids to bed. I could see that he thought I was the parent who took care of all of that. He looked right at us and saw Kevin as my father, with mainly the earning to do, and me as my mother, with everything else to do—whether I had a job outside the house or not. He didn't see Kevin's health as one element in a big balancing act, didn't understand the organism that we are together. He couldn't see us.

Kevin pushed against the limits of what he was supposed to do, driving, folding laundry, trying to make things as normal as possible for me. We scolded him, but he persisted.

And then he woke up again in the morning dark: pained, feverish, sweating, and very pale.

And so it was back to the eighth floor for a course of IV antibiotics to treat a hospital-acquired infection that he likely picked up on one of the previous visits to the eighth floor.

The doctors said he'd be released on Christmas Eve, which they no doubt considered a benevolent move. I sent my mother back to her own house so that she could get ready for a holiday visit from my brother. It was sleeting and everything was about to close. I picked Kevin up at noon with a prescription for an antibiotic I'd never heard of.

I dropped him off at home—he was not well enough to sit up in the passenger seat—and set off in search of the prescription with both kids in tow. Our regular pharmacy was frenzied before an early closure. Across the din, the pharmacist shouted that they didn't have the stuff. I could try calling a compounding pharmacy to see if they could make it. The children ran around my feet in a visual sea of Christmas candy displays, and I made those calls. One pharmacy said no, and then they all began to close.

In desperation, I asked the surgeon's office to page the surgeon. He was a young man who seemed interested in Kevin, as a few doctors are when confronted with an unusual and complex case, rather than daunted and distant. As my phone started cutting out, he said he thought he could get the hospital's inpatient pharmacy to fill his prescription, even though we'd been discharged. I would have to go back to the hospital. If we couldn't get it, we'd have to take Kevin back to the ER that night to be readmitted.

At 6:30 on Christmas Eve, I was standing in the deserted hospital basement, in the hallway across from the empty portal to the inpatient pharmacy. Low-ceilinged, windowless, dead empty, institutional white, a place not made for patients. The Virgin Mary was nowhere to be seen; nothing communicated that we would be comforted or cared for. There was a bench and an ostentatiously large plastic potted plant.

I rang the bell at the window and waited. Minutes later, a woman appeared behind the Plexiglas. She seemed not to know what I was talking about, closed the window, and disappeared. Minutes passed. The kids, who'd thought they'd be spending the day baking cookies for Santa, were with me, because there was no place else for them. They were not being good. They were done being good hours ago. They were silly with boredom and hunger, as I'd ignored all of our collective biological needs for hours in order to get this prescription filled. They were running and rolling on the floor and I was half-hearted in my attempts to stop them. I tried to think of a song or a distraction, but I was also crying a little and didn't want them to see, wishing not to add that detail to the ways in which I had failed to keep things normal today.

I texted the thread of poet-moms.

> So I'm in the basement of the hospital and there's just me, the kids, and a window where they're supposed to have Kevin's medicine, but she didn't know what I was talking about, and she just closed the window and walked away. Twenty minutes ago. And I'm just sitting here on this bench beside a fake potted plant crying and thinking about whether I should go and scream and pound on this window.

Their responses came pinging instantly, threaded with welcome absurdity (one wishing herself into the hallway so that she could take a vindictive shit in the fake plant's pot) and having simply written where I was, I started to exist again in one moment and one place, started dropping into character.

I told the kids I needed them to sit on the bench and stay there, as if sharpening the situation for them would also sharpen it for the disappeared person behind the glass, behind the wall, where the medicine surely fucking existed. The medicine I would surely fucking get.

I stood and took a step toward the window.

No matter my long training in waiting and not-needing, no matter my inherited skill in smoothing things over, no matter my apparent talent for subordinating myself, I would get what I came for.

I knocked at the glass. She did not appear. I knocked again. I let my knocking grow long, hard, and unreasonable. I built up a steam of language.

She walked up and wordlessly slid the white paper bag through the hole in the window with her holiday-manicured nails. I put my card in the metal tray under the Plexiglas and said, "Thank you."

I rounded up the kids, buckled them into the car, and drove home through whirling snow where I'd put them to bed and stay up late to make it Christmas in the Mommy-Daddy-Baby house, a hard formation I was not yet hard enough to shift.

• • •

In the kitchen, standing, cedars dripping rain outside the window. In the kitchen after a long day reading words on a screen, a long day responding to messages, a long day trying to say things in *nice ways* when what I want to say is something more like, *you don't understand what kind of alone I want to be right now*. When what I want to say is *do you, colleague, do you, student, do you, mother of my child's friend—know Julian of Norwich, an anchoress who wrote the first book in English known for certain to be authored by a woman? Have you read about how, after receiving visions upon a near-death experience, she made an application for enclosure, a request to live in a north-facing cell, that she might fill up with God, hear the congregation's music only through a thick stone wall, come to an understanding?*

In the kitchen, I am humming my own kind of application for enclosure, cutting an onion as an application for enclosure. I say something to the children without realizing what it was, except that it contained the word *don't*.

And then the phone buzzes in my pocket: a message from my mother. A video of the Island, where she is now, her own mother's childhood haunt and final resting place, three ferries from the mainland, far enough from lights to see the Milky Way and north enough to see the northern lights. The lens pans soft blue water, calm, purpling at the end of the day. Peace, gentle waves, the strait empty of container ships and Alaska-bound cruises. Distant mountains. Purpling, pinking at the edges, already flashes of snow. *Wish you were here,* she writes.

SHADE GARDENS

One day I'm in a house, probably not our house, with my mother and a bunch of other people, and there are other children around, which is very unusual and exciting. We've recently moved to town. I am maybe four.

An adult enters the room, and there's some commotion, and I understand that there's a deer outside. Suddenly the indoor weather turns to anticipation as the children's motion flurries toward the door. As I too start running, I feel the excitement and relief of being in a group, the unfamiliar sensation of being carried along. For a moment it's possible that a divide can be bridged: the world of Mom-and-me—in which we move through the day looking for and watching animals, thinking about what they do, in which we swim together through currents of nurture and wonder—and the world of People, the little bits of social life taking place all around us, from which we generally hold ourselves apart. Flowing across the threshold, I feel the scope of my possibilities expanding from home to Outside. I feel the promise of others with whom we can share things.

I flash a glance at my mother and wonder why her face doesn't reflect the excitement, why she almost seems to be saying, *Wait*, but I blow right past her look, through the door into the cold air. I run behind the kids, around the side of the house, thinking *Aren't we going to scare this deer?* The kids climb up on the back bumper of a pickup truck, and I follow, confused. I squeeze in between two boy bodies clinging to the tailgate. I'm little enough that it's hard for me to get up there and cling on.

The deer we've all been rushing to see is bloody and still in the bed of the pickup, and I remember it being a doe, a creature linked to Mom-and-me, a member of the animal families she explains to me, and I remember the particular fine, variegated brown of its fur, the red gash, the shock of its deadness.

I imagined the deer alive. The children cheer her deadness light-heartedly. Men and boys killed something. The deer, her death, is an exploit to be talked about and remarked on as man-and-boy exploits always are. This is not the kind of thing we talk about together, when we talk about deer. I understand the look on my mother's face. I feel the possibility of connection between us and the world of People dissolving.

• • •

Once when my mother was visiting us in the new house just before Paul was born, she handed me a book, old and hardbound. We were sitting in the car, just back from an errand, taking a breath before walking back inside.

She said she'd been cleaning out some things.

"I don't know if you want this," she said, handing me the book. "If you don't, give it back to me. It was my grandmother's."

I never met my great-grandmother. She lived in Vancouver and died when my mother was little. We have some things that were hers: bone china with a gilded edge, dark wood end tables, and a bowl-back mandolin that she played while singing on the radio. She had five children and ran a big house, but she had no memory of her own mother. Before she died, she asked her children's forgiveness: "I didn't know how to be a mother, never having had one myself."

"When I was on my own, in camp, raising you two," my own mother said from the driver's seat, "I would think of her just doing it on her own, you know, not knowing what to do."

Years later, in the summer when Kevin had finally recovered from his abdominal surgery and hospital-acquired infection, I couldn't shake the coughing, chest pain, and exhaustion that I picked up during the months of his illness. I finally got a chest X-ray that showed a white, ghostlike smear close to my heart.

The antibiotics didn't help right away. I woke up gasping for air in the downstairs bedroom where I went to keep my hacking and sweating away from Kevin and the kids. It was dark and I was scared and before I fell back onto my pillow, I thought, *This is what it's like not to have enough breath*. Paul was still little. I was supposed to be the global comfort to his sweet body, but I couldn't stop convulsing, couldn't comfort myself for very long.

When my mother called, I'd tell her I was *fine*. To be less than fine would disrupt the daily chain of meals and baths and bedtimes. To be less than fine would mean the shame of being *overtired*. To be less than fine would mean burdening my already challenged partner by needing him to be the well person he couldn't be. And in fact I didn't even know how to name the relief I needed. The circumstances didn't allow me to ask for it.

In the dark early morning, I woke up to Paul's crying upstairs. I climbed up to soothe him and put him back down before descending to my dark, cool room. I wandered toward a corner of the basement that we'd intended for reading and writing: a shelf full of books and a small desk. On the desk sat a crimson clothbound volume, *The Mothers' Book*.

When Paul was napping later that afternoon, I opened its oak-and-acorn embossed cover to random pages, looking for the feeling of vertical integration, knowledge passed down, something from the past that I could keep.

The Mothers' Book: Suggestions Regarding the Mental and Moral Development of Children is an anthology edited by Caroline Benedict Burrell and first published in 1909 by The University Society, Inc. The version we have is a fancy one, with a cloth cover and art nouveau design. A child, probably me, has scrawled in pencil on the title page. A single blank sheet in the back of the book shows more scribbling along with a few sums in pencil, in a hand that's not my mother's. The sums are between two hundred and eight hundred, as if a woman was budgeting on a spare page, perhaps inspired by one of the book's lessons in household economy.

It's a manual for raising children to adulthood. In a table, it maps the child's stages in ways that are sometimes recognizable and sometimes antique, showing, for example, how much sleep a child should get but also the age at which a boy can harness a team of horses. The book promises that "by studying it a mother may learn to deal intelligently, rather than at haphazard, with her growing girl or boy."

The manual was a collection of principles for raising not just children but also mothers, training them up in the way they should go. "Expert" advice was supposed to lift the family out of contingency.

As I thumbed through, reading pages and paragraphs here and there, I saw my mother and her mother, too. The never-yelling. The trust they put in children to take responsibility for increasingly important chores. My mother's belief that if she was a gentle and wise enough example, performing grace and extraordinary forbearance, we would turn out fine by following. And maybe the hint of superiority, too, in the sense that this self-sacrificing gentleness, this ultimate patience, was better than what other people did.

I had often felt and disliked this inherited pressure. But I could see the good in it, too. I'm not sad that I was raised by gentle people.

I picked my way slowly through the book, from front to back and then into the middle, following the sense of us as though I were holding a string in the dark.

The index under H:

Habits
Handicraft, the child and
Health, care of
Heroism
Holidays
Home, the child's
Home, the homelike
Home, pleasures at

Home study
Honesty
Honor, official
Honor, sense of personal
Humor

Here: portable notions of virtue, concrete activities, official observances. Here: aspects of *home,* a big and important topic. Note that home exists for children, that it must be *homelike* and pleasant. Note the implied possibility of it being an *un*homelike, failed place. *The Mothers' Book* was read by my great-grandmother in Canada and carried by my grandmother into the United States, two settler-colonial countries. In such places, the settler home is precarious, and by women's labor and sacrifice it must be maintained.

Home: I could not even find it in my own body.

Something about the anthology's authoritative tone reminds me of the parenting articles of now, circulated on social media, in which we're reminded to "ride out" tantrums with love and patience and proximity, with no mention of the fact that we may not have the forty-five spare minutes to sit "hand-in-hand" with our child as she melts down. In which we're counseled never to put a boundary around our time or space. In which we're reminded that if the child is awake during the night, it's okay, because we can "nap during the day" and "it won't last forever." In which we're given admonishments that would be perfectly reasonable if we were not human beings with basic needs and responsibilities and deadlines. This is a discourse that assumes that being a wise and patient parent is in fact our only job, that living for years with severely inadequate rest has no consequence for health or children. It's advice that heals no fractures. It does not ask the systems in which we're lodged to change.

But while sources of contemporary parenting advice generally leave their politics unmentioned and therefore unexamined, *The*

Mothers' Book, beyond the growth charts and holiday craft suggestions, has a thick back section that situates its maxims inside a field of political thinking. And here the sense of *us* I'd been following, indeed an inherited thing, became a hard figure: something passed down to me, but not something I could keep.

Say you're embarking on an investigation of *home,* what it should be and how you might keep it, even you, drawn so often elsewhere.

In a section called "The Child's Home," Burrell describes the mother's all-important role:

> It is the mother more than the father who sets the keynote of family life. She is there when the father is necessarily away; usually, too, it is she who decides the small matters of training, and it is her disposition which determines whether the home shall be gay or sober, full of the dull spirit of work, or bright with the air of interest and amusement. That it is difficult for an overworked mother—and what mother of small children is not overworked?—to maintain the highest ideals of personal conduct for herself and her family, there can be not the smallest doubt; but that this responsibility is hers must be admitted.

In the mirror of this passage, I could see the standard that constructed my feelings of failure and absence. I could see myself, the *overworked mother,* grading papers before dawn and short with the tiny person interrupting me. I could see my mother in a posture of overt concern for me and maybe tacit disapproval at my efforts. I could see the inherited idea of what we're supposed to be: creators of a bright and happy environment, despite being *overworked.* I could see my own mostly tacit shame at my condition. I could see the seed of my persistent absence: the slippery mental entity I'd learned to understand as my self: rising away from the shameful self that needed.

Looking for some other definition, some other understanding, some counterdefinition, you might look toward Helen Hunt Jackson's discourse on "The Homelike Home." But she too develops an edifice. She paints a wholesome, joyous home as the antidote to all social ills, "the prevalence of excessive and improper amusements, clubhouses, billiard-rooms, theaters, and so forth . . ." If men spend too much time in billiard halls, it must be a failure of homemaking, for if home were made well enough, full of "wholesomer joys," men would not be so attracted to vice. She posits that women homemakers, lacking real political power, can control their fates through creative feats of homemaking.

And rather than seeing this role as subordinate, women should glory in it:

> All creators are single-aimed. Never will the painter, sculptor, writer, lose sight of his art. Even in the intervals of rest and diversion which are necessary to his health and growth, everything he sees ministers to his passion. Consciously or unconsciously, he makes each shape, color, incident, his own; sooner or later it will enter into his work.
>
> So it must be with the woman who will create a home. There is an evil fashion of speech which says it is a narrowing and narrow life that a woman leads who cares only, works only for her husband and children; that it is a higher, more imperative thing that she herself be developed to her utmost. Even so clear and strong a writer as Frances Cobbe, in her otherwise admirable essay on the "Final Cause of Woman," falls into this shallowness of words, and speaks of women who live solely for their families as "adjectives."
>
> In the family relation so many women are nothing more, so many women become even less, that human conception may perhaps be forgiven for losing sight of the truth, the ideal. Yet in women it is hard to forgive it. Thinking clearly, she should see that a creator can never be an adjective; and that a woman who creates and sustains a home, and under whose hands children

grow up to be strong and pure men and women, is a creator, second only to God.

Jackson elevates homemaking and homemakers at the expense of homemakers' full personhood, erasing too the women of lower caste whose labor underpinned the wholesome homes of Jackson's imagination.

Having uncovered this home and homemaker edifice, the dollhouse with the hard plastic figures inside, you might have inklings about who and what is served by it. But to make sure, turn to a chapter credited to Teddy Roosevelt, speaking on behalf of the State.

In his address "To the Delegates to the First International Congress in America on the Welfare of the Child," delivered from the White House in 1908, Roosevelt clarifies not just the importance of the home, but the ways in which women's commitment there is a responsibility to the State, clarifying the ties between the domestic arrangement and the sociopolitical one:

> Every rightly constituted woman or man, if she or he is worth her or his salt, must feel that there is no such ample reward to be found anywhere in life as the reward of children, the reward of a happy family life. . . .
>
> I abhor and condemn the man who fails to recognize all his obligations to the woman who does her duty. But the woman who shirks her duty as wife and mother is just as heartily to be condemned. We despise her as we despise and condemn the soldier who flinches in battle. A good woman, who does full duty, is sacred in our eyes; exactly as the brave and patriotic soldier is to be honored above all other men. But the woman who, whether from cowardice, from selfishness, from having a false and vacuous ideal, shirks her duty as wife and mother, earns her right to our contempt, just as does the man who,

> from any motive, fears to do his duty in battle when the country calls him. Because we so admire the good woman, the unselfish woman, the farsighted woman, we have scant patience with her unworthy sister who fears to do her duty; exactly as, for the very reason that we respect a man who does his duty honestly and fairly in politics, who works hard at his business, who in time of national need does his duty as a soldier, we scorn his brother who idles when he should work, who is a bad husband, a bad father, who does his duty ill in the family or toward the state, who fears to do the work of a soldier if the time comes when a soldier's work is needed.

Child-rearing is the site of women's valor, as business, war, and politics are men's; all ideals that might take one's attention from the raising of children are "false and vacuous." This is why our attention must not be drawn elsewhere: so that white men could continue doing what the growing empire required.

My mother kept beautiful houses when she finally stopped living in tents. She had a knack for composing spaces, arranging handed-down or yard-sale furniture into warm configurations: an antique couch, a handwoven blanket, and the lamp I learned to read by; a rhythm of chores and carefully curated leisure; the TV segregated from the flow of conversation. A house free of extraneous noise and clutter. A house in which my friends, if they came over, looked around, sensing its difference. A house in which we didn't get too wild, and news of the world's cruelties, of the boys who described exactly what happened when they microwaved their cat, of the kids at school who burst into tears at random because their parents had been arrested and jailed, the general atmosphere of danger outside—were met with her abstractions, utterances that distanced us from those around us.

I can picture the carved curves of Great-Grandma Susie's furniture in these rooms and the nineteenth-century molding that my mother sanded down and painted, and out in the long yard, the shade garden of plants that could thrive in a deep canyon's limited light.

These places were my refuge. They allowed me to imagine finding a life somewhere else, among some imaginary kind of people. They were my mother's field of creativity. They asked nothing outside to change.

The self-sacrificing woman who tends the impossible imperative of *home* and lacks agency outside of it, the man who fulfills his military and economic duty to the State: hand in hand, these two and their children are supposed to enjoy "better cheer, wholesomer joys," never mind what's happening outside their house, never mind those whose land sustains their industry, never mind what befalls families whose misfortune or poverty *The Mothers' Book* equates with lack of virtue. This man and this woman: I see them still in our "good neighborhood," at PTA meetings and on parent discussion boards, fighting for "walkable neighborhoods," fighting for "safety," raising funds for their children's seemingly endless enrichment, defending their position like a fort, as if they've extended this logic of the *home-like home* to only as far as the boundaries of our small *community*. This good neighborhood, where it's maybe not a coincidence that most of the moms and virtually none of the dads stay home, where it's common to see not a single male name on the list of parent volunteers in Clementine's classroom, where traditional gender roles seem surprisingly intact.

I came to *The Mothers' Book* looking for a precedent. I found it, alongside the grace, gentleness, and forbearance that I recognized in my own mother: motherhood as creative self-sacrifice, motherhood synonymous with homemaking within the private sphere of house and children, motherhood as a natural state fraught with potential to fail, and in failing, sin against one's nature.

I sought an essence—something elemental I could keep and use and pass on, a kind of gold. I found an entanglement: inheritance and disconnection, comfort and self-displacement.

• • •

After Baker, we came out of camp for good. But our story didn't change as much as it might have. We did not find another way to understand ourselves. Instead of tents and trailers, we encamped in houses, making a temporary life in places formed in one context and lapsed into another, places leaning toward ghostliness, places that were easy to leave.

After Baker, we moved to Idaho's Silver Valley, an area that has yielded close to $3 billion worth of metal in the time it's been mined. Dad had taken a job as superintendent of a classic American silver mine, the Lucky Friday. His employer was a stable company whose roots in the region went back a hundred years. For him, the long-sought dream of self-employment ended there, in a mine whose iconic profile decorated the company hats and shirts and mugs our house gradually collected.

As we left Baker, I chose not to imagine what it would mean to leave the house, the yard, the town for a setting I couldn't envision. I didn't know how to arrange myself in the story of a new place. I decided not to think about it until we were in the car heading north.

Traveling eastbound on I-90 from Spokane toward the Montana border, there's a string of old mining towns hanging on in a slim canyon called the Silver Valley, where winter days brighten late and darken early. Kellogg, once home of the ASARCO smelter and huge piles of powdery black slag, flies by first. In the eighties, it tried to remake itself as a Bavarian ski town, and you might still catch a glimpse of an inexplicably German-style 7–Eleven. Past Kellogg stands a larger-than-life miner fashioned from black metal. He is drilling into the empty air above him, a memorial to the ninety who died in the Sunshine Mine fire in 1972: a miner rather than a family man, the men and their jobs conflated in public memory.

As before, we lived in a short-term temporary house before the longer-term temporary one, this time a regular split-level in a one-street subdivision. We pulled into the driveway as strangers in other people's lives—men in plaid or polo shirts against these tall cliff faces carved out of mountain sides, boys on big wheels flying down driveways, girls and women behind the curtains and out of sight. We arrived there in the afternoon, in the middle of someone else's day.

The out-of-place feeling had a location in my chest, in my throat, where I wanted to be able to breathe without thinking about it. My mother, as if speaking from long experience, told me it would go away.

I started kindergarten there and learned how the out-of-place feelings could come again and again and squeeze out other thoughts. I wanted to read and they wanted to teach me letters I already knew. I learned a host of fears. A pack of loose dogs running the neighborhood tore a kitten to death in front of me. I had to get home from the bus stop by myself, thinking about how to get around boys with their threats and sticks and their BB guns, which they were given at age four. I learned that to be out-of-place was to be vulnerable.

At the end of that year, we moved a few miles east, to Wallace, into a Victorian that had seen its last updates in the 1940s. The house was on the edge of town, up a long gravel drive, with a concrete creek bed on one side and a mountainside on the other. The yard was a long, narrow strip of lawn, and the mountain was for us to roam. When I dream about it now, the days are cloudy, the light blue, the colors sharp.

Wallace claims its old-timeyness in signage and on websites: *historic* Wallace, Idaho. When we lived there, the population sign said 1,200. (Now it's 684.) Then, I-90 slowed to a downtown street with the only stoplight between Seattle and Boston. (Now, the freeway flows above the intact, historic-society-registered downtown buildings.)

The whole town lay within the span of a child's mental map. We could walk to school, to the pool, to the library, to the downtown you-bake pizza shop, to the drugstore, to the grocery store, to a friend's house. At the end of the workday there was a noticeable stream of cars down the freeway from Mullan, the last town before the Montana border, where they were still mining silver from a hole in the earth a hundred years old and more than a mile deep.

The Company's office had been located in Wallace since its founding. The decision-makers and their families had lived in close proximity to the consequences of their decisions, sent their children to local schools. But just after we got there, the Company relocated all the executives to Coeur d'Alene, about an hour away and over a steep mountain pass, making us the only management family in a union town.

I might have felt this more, except that so many of the kids I went to school with were the children of single moms and absent dads, women who stayed and worked at the bank or the IGA or the downtown rock shop after the men moved on to another kind of labor. What made my brother and me most noticeably different, even when the miners went out on strike, was having two parents at home.

We lived in the relics of bygone boom times. We walked to school through late-winter drizzle past ornate nineteenth-century mansions on what used to be a Millionaires' Row. "That one has a ballroom," said the neighbor girl we walked with, of a peeling pink-and-white, ornate hulk of a house. The four-block downtown strip was preserved for tourists. Black-and-white photographs of miners, sheriffs, and unsinkable ladies hung in an old-timey saloon where my brother and I were taken for root beer floats in cowboy boot–shaped glasses.

Lawless Wallace was also known for its drug trade, which some of my classmates' parents were clearly participating in, and the open operation of its brothels, which children talked about, understanding what they were but not what they meant for the women who worked there or the men, many of whom we must have known, who patronized them. The brothels were located adjacent to the police station, and it was known that you could trick-or-treat them, so we did: a red staircase, a red light bulb, and a plain white sign with the word *Luxette* in red cursive letters.

Another set of steps, wide and stone, led to the double doors of the town's Carnegie library. I read the books it held like dispatches from the world beyond, small messengers of a seemingly endless body of knowledge that I was constructing as a kind of salvation. By fourth grade, I read whatever I stumbled on: Robert Louis Stevenson, celebrity biographies, *Anne of Green Gables*, *Little Women*, true crime tales of Ted Bundy and Spokane's South Hill Rapist. I made inferences. I sketched out landscapes and countries and histories and gender relations and sex and fear and dreams from what I found there. At school the teachers gave me extra work, and I read encyclopedias and took notes on index cards and wrote expository reports in a corner on my own. I started rising up, away from where I was, into the random scholarship of a small-town prodigy, forming the sometimes-faulty mental schematic that would carry my physical body away as soon as I was old enough.

The mines, to me then, were mostly names on signs. Whether the veins themselves were tapped out or still producing, the names had an aura of victory. Each one was someone's hoped-for strike, the girl of someone's dreams: the Bonanza, the Homebuilder, the Alice, the Matchless, the Nellie, the Star, the Morning, the Aurora.

The names carried all kinds of dawn, and every June the town got drunk for three days in a festival commemorating some joyous mining event of a hundred years ago. But the artifacts couldn't mask the decline of the 1980s and '90s when falling metals prices, automation, and increasing environmental awareness delivered a combination punch. In a mining downturn, when the town was not flush with money for new libraries and new swimming pools and maybe wouldn't be again, were we taught to take greater pride in mining as heritage, to identify with its most optimistic artifacts as a substitute for the material benefits it used to bring?

We took field trips to Murray, the ghost town up the road, and we rode in a cart down one of the old shafts that was used for silver mine tours. In a tunnel, we learned of the caged canaries that served as live/dead indicators of toxic gas. I remember finding little to latch on to besides the story of that bird. I could imagine its yellowness against the wet black stone, vulnerable as the men's bodies too were vulnerable, trapped as the men could become trapped. A bird wears its color on the outside. It is bright and alive in a place with limited light.

In school or in town, we learned nothing of the 1890s mining wars, wherein dozens of union miners, under fire from mine guards and strikebreakers at the Gem and Frisco mines, united to take control of a tram and rolled a load of lit dynamite into the mill. In the resulting martial law, the miners were arrested and imprisoned. But while incarcerated, they began forming the Western Federation of Miners, which would go on to organize during labor confrontations across the West in the coming decades. Big Bill Haywood, a founder of the Industrial Workers of the World, started as a union miner near Wallace. Our town was rich in labor history, but this is not the history we learned.

For a piece of the 1980s, the high-necked blouses of the 1890s were in fashion, and so was calico, and my mother dressed in Victorian-inflected outfits to go work in a gift store downtown that sold fool's gold nuggets and quartz crystals.

We went to the actual, current, working mine, the Lucky Friday, with Dad, almost never. I have one memory walking up an outdoor metal staircase, looking down through the open spaces in the stairs and seeing piles and piles of sparkling rock. It was probably low-grade lead-silver, but it gleamed and it was easy to imagine it as a great hoard that the mine buildings just sat on top of. He offhandedly squelched our excitement with the facts: just some discard, sparkling but without value.

When we first moved to Wallace, Andy and I and the neighbor girls, whose parents had worked with ours as far back as Leadville, were making the steep mountainside into a home, naming paths and creek beds, rocks, and trees, fashioning "meals" from moss and sap, trekking to the remains of old cabins, unearthing tiny glass medicine bottles, our imagination kindling a world beyond the house. In our conception of it, the mountain gave us everything we needed. We called it *Mama.*

While we were out, my mother stayed in the house, looking out the big back window from the kitchen onto a long shady yard. She has told me that she was deeply, dangerously depressed when we first got to Wallace. I remember only one episode: her hands covering her face as she disappeared into her bedroom when we broke a glass paperweight that she described as "from before I was married."

She almost never yelled, never even showed irritation, but one day when I was in elementary school, her art supplies were laid out on the dining table and we asked one needless question too many. She groaned and stated matter-of-factly that we were the reason she never had time to paint, the reason she couldn't concentrate. When you become a mother, she said, you lose the time and space to do anything on your own.

When I think now about her moments of lashing out—so impactful because they were so rare—I hear her own doubts about how much she could ask for herself. I hear the fracture, not plastered over, not covered up.

She started the rock shop job as a desperate attempt to get out of the house, but she quit when the miners went out on strike because she knew other wives would need the hours more. It might not have been a whole year that she worked there.

In Wallace, maybe it was being the boss's wife that kept her apart from the people around us, kept her there but not intimate, as pleasant as she knew how to be. In the frequent social interchanges that grease small-town life, she didn't let anyone see her.

She grew a shady woodland garden in our narrow strip, but there was never enough light for vegetables. In the house, she arranged Grandma Susie's graceful Edwardian furniture, which she and Dad fixed up a little, not all the way.

I understand now: she was not rooting down in the place where we were, but creating an aesthetic, private space within it, a collection of items she could pick up and move when the time came.

During the summer, she would load up the back of our small station wagon with duffel bags and flats of canned goods from Costco, mount bikes onto the roof rack, pack in Andy and me and the dog—a black Lab temperamentally inferior to the first two, but still loved—and take us all to the Island in a single driving day that began at 4:00 a.m. and ended at 6:00 p.m. with ferries and border waits and a lot of fast food in the middle. We'd spend two or three weeks there swimming and socializing with whomever else Grandma happened to be hosting. Dad might or might not join us for a week if he could take the time off.

In 1984, when I was six, in a letter to Grandma, my mother described the feeling of returning to Wallace from the Island, the golden time, the apotheosis of Elsewhere, a point of comparison to which everything else was bound to pale.

> *The Island is beginning to fade away from us as our own daily life seeps in and takes over. It's always a little sad, like trying to recapture a vivid dream that is harder to remember*

with each day. Life here seems dull in comparison until our eyes adjust back to the mountain air. I'm just now blinking the last bit of sea mist out.

But you know the best thing about getting away from here for a bit is that I can manage to bump George off his track a little bit. The first night home we steamed clams and jabbered at him about sunny days on the beach. I don't think he hardly noticed. The second night we had a cocktail hour—little clam hors d'oeuvres and more talk—he softened. But when I really [k]new he was derailed was when he came across the pictures of Humphrey & Claire's boat for the second or third time. It dawned on him that life was pretty darn short to spend very much of it in a situation as bad as the one we are in now. He said to me "If you went back to school and became a chiropractor we could be outta here in seven years."

The labor negotiations are calm for the moment. The strike may come at the end of this month or six months down the road. It seems that it will come and it seems that it will be rough. Little old gentle me raised in the quiet of the city life. I am amazed at the ugliness and unreasoning violence that has so quickly surfaced in our country community. You know that from experience too though. The smaller the community, the stronger the passions during conflict.

This is the how the letter leaves her: longing again for adventure, Dad having settled on a conventional path, both of them with persistent fantasies of freedom. She paints the world with totalizing impulses: rural life is *this way,* city life is *that way.*

She wrote as an observer who sympathized with workers, referring to herself as a company wife with *subversive tendencies*. But she was also surprised by their anger, assuming that reason rather than leverage would change their circumstances.

And in the very act of putting words to it, reporting it, she communicates and maintains distance from the network of circumstances around her, attempting the impossible task of making home from that high and outside vantage point, retreating again toward a limitless horizon.

The spring when I was finishing fifth grade in Wallace and playing softball and wearing miniskirts and neon spandex shorts with neon tank tops and occupying my imagination with the drama in my small friend group and running out of town up into Burke Canyon by myself, for something to do, to hear the sound of my own breath and feet, my dad was called up from the Friday to the corporate office in Coeur d'Alene. We packed up the Wallace house and moved over the mountains. It was an hour and a half away but seemed like a world apart.

Though the office was on the edge of Coeur d'Alene, a town with its share of old houses and street trees and new subdivisions, my parents chose a little strip of road very far away from anywhere else, a one-street subdivision at the edge of a slowly industrializing prairie. The entrance was marked by a gun club. After living in a company town with a sense of its history—at least the last hundred years of it—where you could hear and see evidence of the mines in the roadside piles of mine waste used to sand roads in winter, in the line of pickups rolling into town at quitting time, it seemed that we'd moved into a land disconnected from any kind of purpose besides the holding down of a fort in the mind.

We were technically moving closer to the markers of population density, things like movie theaters and malls and fast food. There were suddenly many things to buy, but there was again no name for where we were. We described it as *outside of* any of several small towns.

*New Yorker*s, *Atlantic*s, and *Harper's* arrived in the mailbox now, bringing news and perspectives from what I began to think of as the real world, meaning the-world-not-here. I lay on the couch under a big window, looking up at the ponderosa pines, and struggled to read these magazines, piecing together an ideal of Elsewhere from academic references whose native purpose and situation I could only guess at.

My father was more gone than ever. My mother would drive across the long prairie to the one grocery store where you could buy organic milk. She worked in the vegetable garden and grew flowers on our large lot; tended pets; read books; watched *Oprah*, then the local news. She became a certified master gardener and gathered great piles of recycling in the garage, to be driven a long distance to the area's only recycler every few months. She listened to a sober-voiced man read a novel every evening on the local public radio station while cooking dinner. She lived in her private world, her own terrarium.

Meanwhile, outside of it, school did not reinforce the lexicon and worldview I was trying to absorb via magazine subscription. We spent something like six months painstakingly reading *The Good Earth* with a teacher whose mission seemed to be convincing us that "other cultures" were different from "ours" and terrifying, depriving us of a critical lens we might have turned on the inequities that structured nearly every aspect of our own lives. We read *Animal Farm* to teach us that that well-intentioned collectivism inevitably leads to authoritarian mind control. (In the early nineties, post–Berlin Wall, this Cold War curriculum had lost its reason for being but continued nonetheless.) We read *Lord of the Flies* in order to absorb the necessity of rules and order. We were supposed to understand that we were very lucky to live when and where we did, to be Americans. And if we wanted to serve the country they were teaching us to revere, the military recruiter was conveniently located right down the hall.

Meanwhile, I remember instances of a prevailing cruelty that no one apparently questioned: a girl, skinny and notoriously mouthy, being thrown down the bleachers by a crowd of boys, and the teacher on duty seeing it, turning her back, and walking away as the girl howled on the floor, bleeding. The quiet shaming of the solemn, pregnant fourteen-year-olds on the school bus, their bellies emerging from the frightening tininess of their frames. Girls wounded by their brothers' knives and guns. The sixth graders who were regarded by adults as sluts rather than children. The neighbor who shot cocker spaniels if their owner wasn't home during the daytime. The boys who shot at one another. The tacit prohibition against talking about any this, against disclosing any feeling that might render you vulnerable. The assumption of nearly every authoritative adult that people, no matter how vulnerable, are squarely responsible for their own pain, that it's a result of poor choices or problems with character. A pervasive philosophical indifference to the idea of help, to the idea that one's circle of concern might extend beyond self and (at best) immediate family. The pervasive belief that people get what they deserve.

The world outside our house seemed not just difficult to enter, but incommensurable and frightening.

My mother saw none of the prevailing logic I walked into every day and could therefore neither effectively counter it nor exist within it. At home, she sliced apples to gently place between the teeth of our pet rabbit, the one we saved from a meat-rabbit pen, whom she then sensitively cared for and gradually liberated from his enclosure, so that the full extent of his personality was allowed to blossom. She failed to recognize what was painful, and sometimes what was dangerous.

When things seemed intolerable, I did the thing I knew: I started working toward another setting for myself. It was a posture of immense and oblivious entitlement in a community where only a handful planned on college, to assume that I would leave and find a place, conversation, and community where I could listen and contribute. It was a posture that saved me.

I kept myself together by racking up letters on the report card and numbers on standardized tests. (Thanks for the vocabulary, *Harper's*.) I haunted the library and, following breadcrumb phrases like "banality of evil" in *The New Yorker*, read Hannah Arendt and Malcolm X and Sherman Alexie, whose local connection thrilled me. I became interested in resistance movements, especially in the role of women within them, something I had never seen and could barely imagine. From the academically favored tone of detachment, I wrote my many successful college application essays on them. But I never thought of resistance or community as something that could happen where I was. It was an idea for later, for a different context.

• • •

The markets for silver and gold rise and fall depending on what else rises and falls, making waves in a sea of gambles. Some investors buy metals—maybe in the form of funds, in the form of companies, so that they never hold the actual stuff—to hang onto value in case of inflation. Money flows into metals when there's no other safe place for it to go. The thinking is, they're rare, they're elemental, there's a baseline level of industrial demand for them, and they'll hold some value when dollars get cheap.

Gold is transnational. Its price doesn't depend on the stability of a single government. You can carry it over borders in your pocket.

This trading makes a shape, far from the holes and toxic ponds that mining leaves in the earth. It's the shape of travelers in airports, in meetings, in conference calls. Some bodies are harmed and worse, their fates determined, by the pursuit of metal—and other bodies, fewer bodies, generally white and male bodies, travel on a web of contrails, connecting to networks, pinging information back and forth, growing wealthy on abstractions.

My parents have touched down temporarily and taken off again, like leaves on a windy day, in one not-quite-home and then another. They are light in their dwellings.

In 2000, my father left the hundred-year-old company he'd worked for during the years of our schooling and hired on with a small, newer outfit based in Vancouver, B.C. It was a risky move. Rather than live in the city in a density they weren't used to and at a cost they couldn't fathom, they rented a condo in Point Roberts, Washington, a tiny blip of U.S. land tacked onto Canada as the resolution of an old international dispute. To get there from Seattle, you have to cross into Canada and travel through the fields of the Fraser Delta to get to the border crossing into Point Roberts.

Once you're in Point Roberts, it's an eerie place. There are a few houses, several condo complexes, and a large grocery store called the International Marketplace, which at that time featured fruits and vegetables well past their appeal scattered around a cold core of frozen dinners. Point Roberts hosted a few professional athletes and executives seeking a U.S. tax address, summertime Canadian vacationers, and a small population of permanent residents. A scattering of bars and restaurants persisted through the seasons, and besides that, gas stations, one post office, and a lot of empty houses.

I lived in Seattle then, having made it to college in the cloudy, green-gray Elsewhere of my childhood dreams, and I visited Point Roberts regularly, happy not to make the drive back to Idaho. The condo was small but the sliding doors opened to the water and at night, the moon. It was sparsely furnished with the landlord's white wicker furniture and pastel beach décor. Morning sun made bright shapes that moved across the white walls.

The apartment was Dad's launching pad for his predawn commute into the city, and increasingly, weekslong international business travel, and it contained the few things he needed and liked to come home to: several bikes in the closet, a barbecue in the back, a little table and chairs where one or two people could sit outside at the end of the day. The drawers and corners collected his accumulations of foreign currency and Ziploc bags of personal care items a person might take on a short trip. He wasn't there much, and neither was my mother, once she began spending time on the Island seeing Grandma through a long decline.

The Point Roberts condo, its permanent state of vacation, was the bargain the two of them kept making with the world.

• • •

In the days when my mother was writing from camp, her letters were addressed sometimes to Portland and sometimes, after Grandma's retirement, to the Island. Grandma had first visited the Island as a teenager, with a friend whose family had a house there. In the early seventies, she stopped there during a guided kayak tour and was inspired to buy an east-facing waterfront lot with money from her divorce settlement.

The land would have been cheap then, when the ferry only carried a few cars. Some architecture students who were friends of my aunt's designed and built a small, ocean-facing summer house with massive driftwood posts and a chimney and hearth made of beachrock, with urchin shells sculpted into it. The materials were modest and salvaged, the view was world-class, and the floor plan was made for sitting and visiting.

The Island is still hard to get to. When people arrive there, they say, "I don't know what's wrong with me. I'm going to take a nap. I never nap! But I'm just so sleepy."

Which pressures are absent? What cognitive demands does the Island remove? What absence thrives there?

The Island has about ten miles of paved road and few power lines. It has a simple co-op grocery, a few little shops, a few little food trucks, and a couple of restaurants. Every business is micro-small and local, usually the work of a sole proprietor, with a couple of notable cooperatives. Through some miracle of local control, the Island will remain more or less undeveloped unless there's broad local agreement to change it. Except at the entrance to the provincial park, the signage is entirely handmade. Because it's no small thing to get a cement truck there, there's very little concrete on the Island, even in the foundations of houses, which rest on log piers.

At night it is very dark. Last time I stayed there, the night was so quiet that I could lie in bed and hear the miniscule blip of koi surfacing on a pond outside the window. The sound of a bicycle rolling down the road tingled my spine.

On the Island, you might see a hobbitlike man riding down the street on a pony. You might see the sea as an immense blue body of water sparkling through an impossibly northern stand of oaks. You might see a cruise ship headed for Alaska, lit up at night like a diagram of itself. You might see three kids on the beach, their parents far-off specks, and, hearing their American accents, ask where they're from, to which they reply, "Sometimes San Francisco. Sometimes Dubai." You might see yachts and sailboats parked in a wide bay of pale green-gray sand.

To get there we travel for two days, through border lines, ferry lines, American traffic, Canadian traffic. We cross barrier after barrier, skip over distance after distance to reach the Island's dream of proximity, where everything there is, is close at hand.

I can ride, walk, or hitch to the community gathering space, where only one movie will be playing, and sit, hot in an uncomfortable seat, with everyone else, knowing that I have no other option, entertainment-wise, and whether the movie is good or bad, regret will not occur to me. I remember the movies I saw there almost photographically.

But however this place opens me, whoever I can be there that I can't be "at home," I do not live there, and there's no scenario in which I could. The flow of moneyed visitors can support only so many people selling hand-beaded jewelry, or handspun wool, or pottery, or paintings, or floral arrangements. The internet's not great. The place is incompatible with having a regular income, except for a few, like the intermittent doctor or the few employees of the co-op where people buy groceries. And my citizenship is wrong.

The feelings enabled by the Island will start to fade within hours of leaving it, along with plans to bike-commute and commitments to get more sleep and see friends more often.

Once I read a book of the Island's history that suggested it didn't historically sustain, from its own waters and soil, a permanent population. The Puntledge camped and harvested seasonally on the shores; it was one of several places within their homeland, as it is for the K'ómoks now. A few early settlers who farmed relied on seasonal off-island income to break even.

Just knowing it exists, having its patterns in my neurons, the name of it connected to the smell of its grasses and its hay and azure and salal color palette, and its strangely charged air, unsettles me by naming a life that I can't live, that is not mine, like another country's coin that you keep but can't spend.

When my parents sold the last house in Idaho, they didn't know where they'd live. They had a small household's worth of stuff and no place to put it. They had the rental condo in Point Roberts, but even their modest accumulations wouldn't fit there. Dad was generally traveling. Mom was often on the Island, still caring for her own mother. So they bought a house there—the only one that was for sale—and had the movers bring the trucks across all the ferries, through a hand-built gate, past a pond, and up the stone steps of their unusual dwelling.

Its builder specialized in a Nordic Romantic style that incorporated local driftwood, leaded glass windows, steeply angled roofs, and ornate trim. As a child I'd been driven to it, just to see its strangeness. It looks like a piece of a fairy tale.

Inside, its shelves kept the artifacts of my mother's homemaking—pieces she managed to hold on to in cabins, and trailers, and house after house. A handmade earthenware bowl from Portland, which she'd been given as a wedding present, in which she'd risen loaf after loaf of bread. An antique cabinet for cooling pies she'd found at a garage sale in Baker that holds her grandmother's china. The enamel plates and cups that my parents still use as reminders of camp. Snowshoe chairs they bought when they were leaving Alaska.

These are the objects of her terrarium, the enclosed world she makes and remakes. I never return without wanting to join it permanently, to feel like it is also my home. I never return without feeling the impossibility of joining it permanently, of feeling like it can never be my home.

In one of the first smoke-laden summers, after I'd been reading my mother's letters and making notes that would become this book, we visited my parents on the Island, joining them in their vision of the good life for a couple of weeks.

Having risen from sleep into the Island's dream of connection, the children wanting the sea, we dressed quickly and stumbled down the road to a piece of beach. The sky was thick with burning Elsewhere, strangely pink.

The children climbed across huge log piles. I knew from being a child there myself how you have to pay attention with your whole body, with feet and posture, to the particular contour, the particular lean, the balance point and the stone beneath. I paid attention, too: the way the tides arranged worn-down oyster shells, the weathering of wood, the shifting scent of a sea lion's desiccated corpse, the complexity of the geologic zones that meet there, the mixing and exchange of matter that each of us will one day join after becoming nameless.

In the years when we were growing up, when my mother mostly could not steer herself, we floated with her in many places where there was so much quiet, like here.

My mother arrived on the beach a few minutes behind us. She emerged lanky-aging, freckled-graceful, her short hair the color of cedar bark, quietly reveling in just being there. I could feel the ways in which I belonged to her and always would, belonged to the body that generated and carried me from place to place and taught me what it knew, taught me, in the best moments, how to read, how to pay attention. She is the first home, not a state of grace, but a human being, unsettled and imperfect and perfect too.

• • •

There are various ways to estimate the amount of gold that exists above ground. If it could be gathered to one place and melted and formed into a single object, it might be as small as a sixty-seven-meter cube, approximately the height of a giant sequoia.

The earliest memories I have, the terms on which I formed, were in the midst of a search for precious metals. Silver and gold are valuable because they shine, because they're difficult to extract, and because they are scarce. They're the gleaming, essential elements.

Now my time and energy are spent in the middle of—in the service of—things that are difficult to value, endeavors whose yield is struggle and joy. Beings and experiences that are plentiful, beings and experiences that radiate beauty and worth and still can't sell their time for a decent price, beings and experiences more understandable and beautiful within their context than removed. Poets and poems. Reading and study. Students. Children.

A different currency is required.

The summer after Kevin's hospitalization, after the pneumonia diagnosis and finding *The Mothers' Book* in the basement, I couldn't get well. Well enough to stand, but not well enough to move much. Soft and pale from being inside, still coughing a never-ending fatigue from my lungs. I managed to get the kids to preschool and back and answer my email, which never stopped. The garden grew wild from my inability to weed or trim, the rough lawn long and blooming with dandelions.

For months, we'd been planning a trip to Chicago to visit old friends, just Kev and me, for five days. We decided not to cancel, and Mom and Dad came to stay with the kids. We packed and left, light and unburdened, alone for the first time in five years.

In Chicago, we stayed in our friends' apartment, enjoying the company of their daughter, the same age as Clementine. We planned our days around the heat and humidity, leaving and returning to the apartment early, swimming before the lakeshore blazed, bowling in the air-conditioned afternoon before long naps for me and reading and guitar breaks for Kevin, and late nights of wildly associative conversation and tear-rolling laughter. In Chicago, no one needed us for anything. Our hosts prepared our meals and insisted that I rest. In rooms of beautiful found and hand-fashioned things and with good, strange books, I rested as I never could in my own house. In a city that was not my city and a home that was not my home, beloved people healed me from the lungs out.

• • •

I have imagined Clementine reading this book. She is a different person now than she was when I started, taller and heavier, the landscape of her mind more private. She is more sure of herself and less sweet.

A little way back in this dark space where words cross time, a blip in my life but long in hers, I started writing to her directly. In this letter from the winter of 2018, when she was six going on seven, there's a string of days tied together, from a time when I'd made myself some time. It's hard to imagine them as written toward the person she is now. But in them I am trying to say, for her and for myself, what it means to be here. They're to her and they're also because of her.

Dear Clementine,

I'm finally alone today. You and your brother have been home from school for most of this week for snow days. We never know when the snow is coming in Portland, so we get inside and stay to watch the streets go white. If a storm catches us out in this city without plows or salt, we might not get home—or get to you—for hours.

I'm going to pick up the house again and light a fire and watch the spare flakes floating down outside in the record cold. It's late February. Blooming branches of daphne bend under four inches of snow. This is not typical.

You're growing up in strange weather. Things bloom at the wrong time and the ocean is full of jellyfish I never had to fear. Last summer the sky rained ash for days, and you and Paul saw me crying. Hadn't they told us about this in school? That the warming would bring more frequent fires? Words alone didn't get across what it would feel like: *for weeks in the summer, your children won't be able to breathe.*

On that day, when we woke up to ash covering our cars like snow, the Justice Department had just announced another unjust decision against a group of people whose parents carried them across a border as children. They're in danger now, people I know and care about, and I know from history that a government that invents reasons for people to fear their most vulnerable neighbors, that picks and chooses when to care about "law and order," endangers everyone eventually.

I went to a rally that evening downtown and saw some of my students there. Your dad fed you and put you to bed. I came home to a drawing you'd made with black marker on black paper. The words, which I could just make out, were surrounded by moons and stars: *Mom Powre.* You, the fact of you and the way you are, have pushed me to grow my *Mom Powre. Mom Powre* is for you. The one noun must not contradict or work against the other.

You're growing up in a landscape that will keep catching all of us off guard. The knowledge I can give you won't always match your circumstances. What my hands know how to make won't always be the things you need. You won't be able to rely on ideas and ways of being that might have served in different times.

The earth is dynamic, but it's not wrecked. And, just like in the days after you were born, I still want to tell you everything I know.

This morning I stood in front of the bathroom mirror and combed your thick, straight, pecan-colored hair. You lifted your neck and told me how your hair wanted to sit, and I laid each strand the way you wanted it to go. I greased your shoulders with lotion to stop your dry skin from itching. You stared intently at my face and said, "Mom, do you know what's true?"

"What."

"Your face looks almost just like a heart."

In the past years, I've been reading and writing about my mom's—your grandma's—letters from camp. I set this one aside at first. But there's something in it I haven't forgotten.

December 1977

I tried to take a picture this morning of the snow outside but the polaroid got stuck in the cold so I had to take this shot thru my dirty window. Today is New Years Eve and as you can see in the picture it's a beautiful day. George is out shoveling snow and I'm inside with a lump of sourdough bread and madly getting sheets and baby stuff washed. Most of the time I string up the laundry around the house but the sheets are too big to do that so on nice days I can hang them outside. Only problem is nice days are often few and far between.

George and I are finally feeling quite relaxed and confident about the birth process. The classes and Bradley book have been a real training session and now we can both hardly wait. It's still a little hard to believe that the end result of all this is an actual person—our son or daughter. That part I'm not so confident about! I saw the doctor yesterday and he says altho the baby has dropped it can easily be another three weeks. So I guess you had better hang in there for a while longer. Can you still get the discount price if you let them know a couple of days in advance when you want to leave? I figure if we let you know when I start to go into labor you could come two days after that because I have a 24 hour minimum hospital stay anyway (baby has to be on oxygen that long). And you don't have to be here the day I get out of the hospital either—I imagine Geo and I could use a day alone to get used to the three of us.

It's now evening time and Geo is just closing up his shop. I went out to see him and he's got the barrel stove simmering away and he's cutting boards for a cradle by the light of the Coleman lantern.

We declined to go to all the New Years Eve parties tonight and I think we're just going to have some home made Mexican food and play our Portland monopoly game tonight. I'm absolutely exhausted so I doubt we'll make midnight. My blood pressure was up a little last doctor's appt. So I've been trying to get more rest during the days but today I forgot all about it. I made all my decent nighties (two) into something I can nurse in and still look nice in for my hospital stay and tomorrow Geo and I will pack our bags. We have three to pack, mine, the baby's, and the coaches. Geo has to take me, his coach card and notebook, powder, lotion, washcloths, watch, roll of dimes and list of phone numbers.

That winter before I was born, she made a block print of our snow-heavy house. Her hands moved in the material world, sending a handmade picture of our lives to people far from her. It was a product of her vision and her body. She was drawn to a makeshift life by the adventure of shifting but also the pleasure of making. I love the things her hands still make, the pieces of herself that she leaves in the world. And I'm glad she wrote things down. She recorded the existence of these dimes, each one representing a conversation yet-to-be. Her traces—the hot stove, the cradle boards, the cold-jammed camera, the objects she carried on the day of my arrival—anchor me to a world made of wood smoke and sawdust, bed sheets and phone lines, everyday metals, and ice. The fact that she wrote her life at all, allows me to see her story through the lens of my own time, to pick up and consider the artifacts of her life. I can choose what to revise or carry forward, as you will too.

My life is like and not like my mother's. I'm not marooned in a tent or a trailer or a cabin, not so cut off. There are strings of text messages flying between me and several groups of friends. Four active threads right now. If I need advice or encouragement or a joke, if I need to say, "Hey check this out," I pick up the phone, which is not in a mine building or a town hours away—but which I carry in my pocket. I'm trying to make a more connected life for you to grow in.

The ground is cold now, but cover crops are depositing nutrients in our vegetable beds. In the summer, beans will twine up our trellis and you and your brother will turn the woods into one long story, and I will count out the minutes like I do, for writing, for work (even when I'm technically off work), for the dishes and tidying up, for exercise—mine and yours and the dog's—for swimming lessons, for science projects, for play dates, for working in the garden.

At the end of the day I'll lie in bed and go over and over and over all the things I didn't get done.

Maybe this year I'll learn to go through the day on your schedule where we eat when we want to, where I follow you from one imaginary feast into the next.

Rain clouds are gathering now, and I'm looking out the office window at the ruffles the cedars make as the wind comes at us from the south. The trunk solid, the fronds rippling top to bottom.

Do you know what's true? This is something you like to say to me before unreeling a string of facts. *Do you know what's true?* We're resting at the end of a long cascade of consequences that descend from a collection of ideas, views of the world, shapes in the mind. One of these is a tendency to make the multiple single and the complex simple, to remove things from context, the better to profit from them. All of the prosperity that allows me to sit here looking at trees and typing comes from a way of valuing that doesn't recognize bodies or animals or land except as means to an end.

I can't watch you dreaming around the house and forest, talking to trees, embracing your brother, twirling for the sake of twirling, without remembering all the forces, outside and ahead, that will work to separate you from yourself, that will whisper that your beauty can be measured according to a standard, that your body can be ignored, that your time and brain are without value except as the means for someone to accrue profit, that your ability to see and hear and care for others is, at best, worth minimum wage.

At our well-regarded public school, in our *good neighborhood*, you're already learning the art of turning yourself off and completing tasks without purpose or meaning beyond checking boxes for the State. You're gaining the small art of compliance for its own sake.

What I can do to keep you together, no matter the school you go to or the job you work, no matter how old we both get, no matter whom you choose as your partners and companions, I'll do. I will not pass on a gaze that splits you from yourself.

It's Monday and you're at school. Your brother is napping. I am at home on a Monday. I'll have Mondays and Wednesdays off for the next two years. I'm trying to have the thing I'll never have—enough time to do all the things I need to do, be all the things I need to be—and I split my job so that I could be home two days a week. Home with Paul and you in the afternoon.

When you're gone at school, I tidy up, organize things, listen to the radio. I pour the leftover coffee onto the roots of the blueberry bushes outside to give them the acidic soil they need in order to give us many, many blueberries this summer. Home alone I *read*: an hour in the afternoon with a book and not a screen, next to a window with trees outside, just like when I was a kid. Home alone I write and write and write, things I wouldn't have known how to try before.

The cedars are still today, and the bigleaf maples haven't leafed. I can see straight through them to the blue distant hill. I used to feel this wanting feeling when I'd see the shape and color of relative distance: when I lived in Seattle, it was the afternoon sun flashing off mirrored towers across the lake; when I was in Italy, it was a massive church dome in the distance, and in St. Johns it was the blue hill of Forest Park across the Willamette. It was as if I could only locate myself if I was far from a fixed point. From long training, I did not know how to imagine that I too was in a place that mattered. I too *was* a place that mattered.

In an hour or so you'll be home from school. I'll wait in the yard, and the bus will stop, and you'll sulk down the steps, dragging your backpack, books and paper in your hand, your coat hanging off your shoulders. I'll ask you how the day was, and you'll say, "Thumbs sideways," meaning you're ambivalent, or "Terrible!" which is usually because one bad thing ruined the rest. I'll bring you inside and try to talk you into a bath while you unspool a story of who said what to whom. You will scowl at each injustice as you tell the story. Good.

It's possible to see now, as an image on a screen, the shape of underground ore bodies without ever breaking the surface, the formations of minerals and metals tightly layered and laced by the long story of the earth. I am learning to recognize presences rather than endless departures—what *is* rather than what is *never enough*. I am learning to see the kind of home that makes itself around us, depending on to whom we are present, to whom we pay attention, to whom we are honest, to whom we are accountable, to whom we respond. I am trying to learn what it means to be more present and more honest, less smoothed over.

Since I started wandering in the dimension that would become this book, I haven't been much of a homemaker. My cabinets are a mess. Rooms remain unfurnished. The house is a long to-do list that I often ignore, especially as I've turned again and again toward this story of our family, and toward my students, and toward problems we can't pretend to shut the door against. I have failed where my mother excelled, in this elevated domestic sphere that was her means of creativity, her measure of control. If I could pass one thing down to you that I did not myself inherit, it would be things I'm still learning myself: people-keeping instead of housekeeping and a more expansive version of whom you might consider kin, whose concerns you understand as linked to yours.

I have written about the experience of growing, birthing, and raising people as an absence from oneself. I have written about mining as the making of an absence, above ground as below—the absence of context and a failure of story. I have written about the early shape of absence following me into the present as if the early absences of my childhood were inside me. I have sung this sense of *gone*. I have rhymed it with *bone,* the deepest sense of structure. I'm learning how to meet myself again. I'm learning how to write differently than I was trained to, learning to fill some of *gone* in my *bones* with *context*. I'm learning what it means to write more honestly.

Today, after your bath, I'll dry you and brush your hair and dress you in warm clothes. I'll lay back on the couch and if I'm lucky you will find me, and I'll quiet what's restless in myself, and if you want, I'll tell you how close we are to rivers and mountains, tell you where the water comes from and where it goes. For as long as you need me to, I'll be your ground, holding you here and here and here.

EPILOGUE

The Ghost Road & the Grass-Child

1.

After we've been at the cabin for a few days, Paul, just past his sixth birthday, starts fashioning our moments into movie scenes. It's been weeks of lockdown and we're taking the only vacation we can. We're high in the mountains and far from a cell phone signal, electricity, other people. Paul's imagination is blooming, maybe too hard. The kids have been away from other kids a long time, and I have lost sight of what's typical.

One afternoon I'm at the river with him and Clementine. We've spent the hot midday wading and looking at rocks. When it's time to wander back to the cabin on scratched legs and make our peanut butter sandwiches, I gesture at the bank.

"Why don't you two go first. I can lift you." I hoist their bodies over the sharp brush, start clambering up myself.

"This is the part where the mom gets eaten," Paul says, "because she decided to go last."

When the sun finally slips behind the mountain, it's suddenly ninety instead of one hundred, then a merciful eighty. Observing tradition, the adults—my parents, Kevin, and I—drink Moscow mules on the warped deck. The kids rip into little bags of chips. The word *evening* makes sense in a desperate, tactile way: it's a period of balancing the skin and the air, opening up the cabin so it cools off enough for us to sleep in it.

"This is the cocktail hour scene," Paul says. "I said that just now because it's in the script."

The cabin we're staying in has been owned by someone in our family for fifty years. It sits on a parcel of private land with a handful of others like it, in a small valley beside a river. Nearly a hundred years ago, a man dug a series of ponds into a meadow, stemmed the river to fill them, and built a set of rustic dwellings, each with its own postcard view. We are told he intended them as getaways for Hollywood stars.

The man is long gone and the cabins have aged, but the place still offers a curated version of roughing it. There's no electricity, no cell phone signal, and limited access to semi-potable spring water. But the lakes are stocked with trout for kids to catch. A caretaker stays on-site most weekdays and keeps the cabins supplied with firewood, but we have to chop it. If we ask him to, he'll trim shrubs to improve our view of a small waterfall.

Paul is not the first to imagine a movie here. The cabin itself contains a nonfunctioning barrel and bugle, props from *Paint Your Wagon,* the 1969 Western musical that was filmed nearby: an entire old-time town built and abandoned along with a rooster named Broken Toe that wandered the valley for several years after.

Kevin has called these cabins a kind of rustic theme park, trafficking in symbols of scarcity, solitude, and self-reliance for those who wish to escape for a long weekend. My grandfather, a surgeon, bought the place in the middle of his career—even though it took him a day to get here—as a place to set work aside. For an owner or guest, it's a place to pretend, a place that asks little from you, a place you might leave feeling refreshed. On this trip Kevin is having a good time using it as intended, fishing and swimming and not being able to check messages.

But we're staying longer now than a long weekend—longer than a week—and the summer is a good deal hotter than it used to be. For much of the day, we're just trying to endure it. And during a global pandemic, a national uprising, and increasing autocratic threat, there's nothing usual about the business from which we are supposedly taking a break. Maybe you can take a break from electricity and therefore the news, but you can't take a break from what's happening.

During long trips to town for ice, while untangling the children's fishing lines, amid the scenes of rushed cooking in the short minutes when the temperature allows it, I am wary of the children's deepening affection for this place, the games they get up to here, the way they lapse all the way into a fantasy so deep that all of life becomes a play for days at a time with nothing and no one to break the spell. They want to learn to chop the wood and catch fish. They forget their city friends' names. They want us to buy them cowboy boots. By the minute, they are changing.

There's nothing wrong with the sound of the river outside the window. Nothing wrong with having an alpine swimming hole to yourself on a hot day. Nothing wrong with getting to know this dry country's cast of birds and plants and animals. I sleep so well here. I should be grateful. I *am* grateful.

I am also, often, uncomfortable. When I question whether this place is meant to be stayed in so long, my mom reminds me of how she lived here a whole summer once when she was in college, driving all the way to town (ninety minutes away over jarring dirt roads) and back on weekdays for a small job. She loves being far away, cutoff, out-of-touch; the feeling of *nothing else, no one else, just us*. She's an expert on being here, knowing in detail which supplies the cabin needs and doesn't need; where to spend the day on the river; how to bathe without using too much water; the best time to rise; the best time to sleep; what bird that bird is; how to keep the bedrooms just cool enough. Her self is firmest surrounded by this negative space of being without.

Here I can see how her early experiences in this particular contrived austerity may have started her down the path toward a rougher kind of *roughing it*—a kind you can't easily drive away from.

2.

If you want to stage a scene of abandonment, dwindle the other vehicles on a smoothly paved road until there are none but our two. Give the rushing Powder River over to vast stretches of piled rock made by decades of dredging for gold. Let paint peel off the mile markers until they're illegible. Let the sun bleach a *scenic viewpoint* or *historic marker* sign to white nothing.

We—my parents and the dogs in front, Kevin and I following with the kids—enter this forgotten zone on a long day trip from the cabin, searching for Granite and the Cougar Mine, one of the places we used to live.

For the past several years I've been writing about this experience. I've been reading letters my mother wrote to her own mother during the years when we lived in tents, trailers, and cabins. I've been trying to write my way into her words, trying to untangle the knot of fear and absence I carried with me since my first child was born.

In the Granite camp, my own first memories formed, my idea of the world's shape: us, very far away from everything but a mine, for reasons I couldn't understand. The few people around us, permanent strangers.

Now, as we caravan up the scenic highway, I'm waiting for the sensation of *familiar*.

"Where are we? And how long till we get there? And what is going to be there? What will we do?" the children ask from the back seat. *What is this nowhere road? What can I make of it?* Their questions point to the specter of pointlessness.

We pass a tiny but still townish town on the way up the mountain, Sumpter. Let's call it a sub-ghost town. In the fall, it hosts hunters from all over. There, just beside the road, we see the massive dredge that made the miles of tailings piles. It looks like a riverboat attached to a giant metal apparatus for raking up the river. We'd planned on maybe finding something to eat in Sumpter, and in normal times there might be a place, but everything in town is closed because of the pandemic.

By the time we reach Granite, we're on the back side of the mountain and we haven't seen another car for miles. The sunlight feels relentlessly bright, the air relentlessly thin. We round a bend, and there's a large town gate. Age-stained board buildings, log structures, and a few newer dwellings spread up the hillside on our right.

As we pull across the town threshold, I'm wondering if there's going to be anything to see, if the long drive will be any kind of worth it.

In Oregon, ghost towns dot designated *scenic highways*, roads there's little reason to drive except as a sightseer. On these long, just-paved stretches without amenities, you have to be careful not to run out of gas.

Lists of ghost towns worldwide show how often they are artifacts of mining. There's a form to their story: precious metals found, town slapped together fast, a decade or two of rollicking (sometimes violent) business at most, and then the easily accessible ore is gone, or the railroad goes somewhere else, and then the people leave. In the twentieth century, companies mined in the old places when the price of metals was right. But mostly the towns commemorate themselves if they do anything at all: preserve a few old buildings for tourists, assemble a small museum, hawk small wares.

A ghost town is one that persists past its original context. The story that gives it identity is in the past. In this sense, the town itself is a ghost. But you can also imagine it as a town abandoned to the dead.

Some living people still reside in ghost towns. To settle or stay in such a place is to hold yourself apart from so-called normal life, inhabiting its most minimal shape (streets, structures) and not its content (social purpose and economic activity). To settle in such a place is to take yourself away from external reference points that might construct you from the outside. To settle in such a place is to choose a simplified, minimal environment. You wake up with nothing but the hours and what you can make of them.

It may be what my mother seeks in her gravitation to way-out nowhere: our selves rising to the surface in a place where ostensibly nothing else competes.

Even in this sentence I've erased people: "nothing else" in fact means *no one* else, no stories or dispositions that might challenge your own, none of the difficult work of getting along with others.

Granite sells itself to anyone who happens by as a ghost town. It attracted curious visitors when we lived there, too. Now we're the people driving up out of nowhere, to gawk. The bar and grill just along the highway is closed, and for sale. I pull up beside my parents and we talk, driver to driver, through open windows.

"Why don't you guys go ahead. We'll just wait here," my dad says, leaning out the driver's side window. I can see my parents leaning together, conferring, as if maybe they don't agree. I'm imagining that she is curious and wants to look around and he doesn't see the point, but that's just the momentary story I'm telling myself about the forces at work in my parents' partnership. They are a nucleus whose coherence and incoherence is sometimes hard to understand.

We cruise Granite in the car, conscious of being maybe/always in someone's front yard. A hand-painted mural on someone's shop-garage: a woman with flowing hair holds a baby and radiates light. The Little Free Library: a box on a post with a clear door is full to overflowing, and the books are not unlike what we'd see in Portland, a lot of James Patterson, nothing weird. In the center of town: a tiny museum in one of the nineteenth-century buildings, closed; a post marking the miles to Yuma and Honolulu. And a small cemetery, where we decide to stop and get out.

There's a short picket fence around the oldest part, but the graves spill outside of it. The grass is unmown meadow and it has a particular smell, like a tomato plant but sharper. *Familiar*.

We wander the graves, coaching the kids on respect. My daughter silently reads the stones and looks up at me. *Baby, baby, baby*. Children under five. Children under ten. The stones simple or ornate. Family plots for just a few of the nineteenth-century families, the man's name large and everyone else a detail. More recent deaths of old people marked by stones outside the fence; none of them, from what we can tell, born here.

There's a large piece of black granite for the Marshall. We turn around and see that my parents have joined us, to find this marker.

I can hear engines revving over stray male laughter. *Familiar*. Always male.

A large dog comes bounding, and the owner, maybe my age, walks over from a large, corrugated metal shop. I can imagine him: finding this place, bringing money earned elsewhere, a whole town more or less to himself, mountain roads and a four-wheeler and no one to bother you. We're at the bottom of the cemetery, and he and my parents are at the top. I can hear my dad telling him we're looking for the mine.

Paul can't place Granite in his script. In the cemetery he looks up at me earnestly and says, "You're going to have to help me get this place out of my mind."

3.

Fifteen minutes later, we're driving out of town, following my parents up a dirt road veering sharply up into the mountain. *Familiar*: the way it tilts toward the sky, the intensity of the sky against the trees in this thin air.

And then we're out of the cars, stepping around a sign that says ROAD CLOSED TO VEHICLES.

Dad stays in the truck for reasons of limited mobility and generally not wanting to go. "There's not going to be anything there," he warns. "Nothing to see."

But Kevin says, "When you all talk about this place, I want a picture in my mind. I want to see it."

And so we set off up the ghost road. Knee-high meadow grass grows through shallow ruts and hardened bulldozer tracks, and we have to walk around the occasional fallen branch. But otherwise the road is passable on foot. The sun is directly overhead, and we didn't bring water.

The kids are steadily complaining but also refusing to go back to the car and stay with Grandpa. Whatever's ahead of us—and they doubt that it's anything—they don't want to miss it. Sunscreen-less, Paul takes off his shirt and leaves it behind in the grass. I can feel his fatigue and thirst and impatience in my own body.

Neither of my parents is sure we're in the right place. And if we are, no one knows how far up the road the mine, the camp, the old cabin might be. I contemplate what it means not to be able to find—and maybe not to want to find—a place where you lived for three years, a place you brought your baby home to. Twenty years ago, I lived in at least six different Seattle apartments. Some of them were shitholes, all held at least a few good memories, and while I don't know the exact addresses and the city has changed a lot, I could find any one of them easily.

"Doesn't any of this seem familiar?" I walk beside Mom while the kids complain along behind us.

She says none of it does. "The road was dirt for maybe fifteen miles, all the way to Sumpter. In the winter we were responsible for plowing the whole thing."

"Do you think we're in the right place?"

"Well, the guy in the cemetery said you could get to the mine from here, but on a four-wheeler. Dad asked if we could walk in, and he said probably. I don't know what's left up there. The guy said some old equipment. Maybe there's something left of the cabin."

"Does Dad know how long it's been shut down?"

"Right after we left. Forty years. Do *you* recognize any of it?"

"I thought I did. The way the road rose up when we came in." That rising bend that said we were nearing camp: my child mind would make a scene right there, animate our place, learn to love the feeling of coming home even when being there meant days with the same three books in a single snowed-in room.

In one line from her Granite letters, Mom says she's coming out of camp and will need a dress for her sister's wedding: *It's been so long since I bought a dress, I don't even know what to look for anymore.* Once you are that person who lives nowhere, you may have a harder time knowing how to dress for somewhere. You may have a hard time seeing your way out of the scene.

After twenty minutes of walking, the road ahead looks foolish, too steep, about to dwindle into meadow. The kids are not going to make it. I look sideways into the shade of subalpine conifer forest, the storybook trees with space between them, blades of grass rising out of hard dirt into a blue-green haze. This field of shade and white wildflowers. *Familiar.*

I spent long days in this field, or in a field contiguous with this field. I passed whole afternoons inside this tree-shadow, inside my own mind, placing myself and our family inside a less lonely and more dynamic story: orchestrating stuffed-animal ceremonies, imagining Beatrix Potter–inspired societies and etiquettes and customs.

Just looking into it, I can feel the cool and the privacy. In my mind, I can see a small body, shirtless and wearing too-small sandals, running into the grass, her imagination blooming hard.

The kids beg for a break, and we stop. Mom and I agree that the road is fading and we have to turn back.

I walk a little ways off road to squat and pee (*familiar*). I let Paul climb on my back and rest his hungry, thirsty, tired body. Looking over at us, Mom says, "I remember now. Cross-country skiing down here in the winter, with one of you on my back, just like that. This was it. This was definitely it. Another time. Another time when we're more prepared, we'll have to come back."

The dreaming grass-child is with me when we're back on the paved road driving down toward town and the long-awaited lunch. She remembers the drama of a tall rock face we pass, the unlikely castle-ness of it, the strangely beautiful kinnikinnick spilling down it as if a plant could embody the cascade of water or the upward curl of smoke. The grass-child is with me in the park in town, as we all eat burritos in the shade of massive old locust trees.

Back at the cabin during another cocktail hour scene, my mom looks at me. "Granite. What did you think?"

"Pretty far out. Not much there." Though brief, we both know my words refer to an old beef: that she had me and kept me in a strange and narrow place where I couldn't see the space into which I might grow. That I had no way of understanding the experience except the constant uprooting of what she called *adventure*. That even after the various tents and trailers and cabins, the rootlessness only continued. That I couldn't share the meaning she made of our life. That I have come to see it as a series of choices and consequences rather than as an exhilarating test of our ability to make do with very little.

"Well, when we were there, the mine was there, so it did seem like there was more going on." For "going on," a mine was always supposed to be enough. For her, it was.

And my complaint—about their choices and values and the identities we're still supposed to be playing at—is old, a building standing too long after its context. Its timbers are stained with winters of heavy snow, and its child occupant left it long ago to make choices of her own.

The kids' laughter fills the small canyon. They've been clunking around in a pair of cowboy boots made for an adult, taking turns pretending to rope each other. They call it *playing Western.* To pass the hours, they invent mock purposes, tiny objectives, ways to give their bodies meaning. They figure out how to animate our setting, latching on to stories or a set of relations from which they can play with our increasingly uncomfortable and shapeless right-now.

They retreat deeper and deeper into their own shaded meadow. I worry they'll fall in love with the *way out here* of this place. I worry that they'll become people they can only be at a distance, self-narrating their way into isolation. I worry that they'll venture so far in that they won't be able to come out.

But it's resilient, too, as the grass-child reminds me, to keep conjuring your own wider field of action and consequence when you find yourself on one forgotten hillside after another. Maybe we all have to imagine ourselves a little, even when we've stopped noticing the slippage in and out of scene. Maybe this dreaming will get them through until they can play a better game.

SOURCE NOTES

This book has existed in many versions as I drafted and revised over a period of six years. Early versions attempted something closer to a scholarly approach. As I found the story I could tell, I wound up removing a great deal of the research I did, but the early drafts helped me shape stances and positions that remain. Here I want to acknowledge some of source texts whose details made their way into *Mettlework* and some others that helped me develop a knowledge base and perspective.

In writing about geology, I relied on Ellen Morris Bishop's *In Search of Ancient Oregon: A Geological and Natural History* (Timber Press, 2003) and Steven Earle's open-source textbook *Physical Geology, First Edition* (Pressbooks, 2015). The website minedat.org was generally useful for geology and mining information, especially in the section on Leadville. A panel at the 2019 AWP Conference in Portland called "Reclaiming the Mine: Nonfiction Writers Explore Mining as Method and Metaphor," with Byron Aspaas, Ander Monson, Katherine Standefer, and Elissa Washuta, shaped my thinking on mining as metaphor in ways that I want to acknowledge.

The section on Slick Rock contains facts from Daniel R. Shawe's 2011 report for the U.S. Geological Survey "Uranium-Vanadium Deposits of the Slick Rock District, Colorado: U.S. Geological Survey Professional Paper 576-F" as well as Kathy Helms's 2019 Associated

Press article "Mining Camp Alive in Memories of Navajo Uranium Victims."

In writing about Granite and Baker, I used the Baker Heritage Museum and the *Albany Democrat Herald*'s reporting on Bud Morrow's murder as general background sources. For specific mining and geology-related facts, I studied Howard C. Brooks's *A Pictorial History of Gold Mining in the Blue Mountains of Eastern Oregon* (Baker County Historical Society, 2007).

Some of my writing about North Idaho's Silver Valley was inspired and informed by Patricia Hart and Ivar Nelson's *Mining Town: The Photographic Record of T.N. Barnard and Nellie Stockbridge from the Coeur d'Alenes* (University of Washington Press, 1984). I also read the United States Department of the Interior's 1972 *Final Report on the Sunshine Mine Fire Disaster*, though the details of that report didn't make it into the final version of this book.

In my writing about the Island, I refer to a reading of David Gerow and Elizabeth Smith's *Hornby Island: The Ebb and Flow* (Ptarmigan Press, 1988).

My writing about Portland rests on years of reading and teaching about the city and the region that is difficult to untangle. My interest in Portland's history, in contrast with its image, started during my husband Kevin's internship at the city archives in 2008, when he was studying to become a librarian. Notable readings and experiences that informed these sections include Timothy Egan's *The Good Rain: Across Time and Terrain in the Pacific Northwest* (Knopf, 1990), public appearances by Walidah Imarisha during the 2010s, Mitchell S. Jackson's *The Residue Years* and *Survival Math: Notes on an All-American Family*, James J. Kopp's *Eden Within Eden: Oregon's Utopian Heritage* (Oregon State University Press, 2009), David G. Lewis's *Quartux Journal—Critical Indigenous Perspectives*, Gray H. Whaley's *Oregon and the Collapse of Illahee: U.S. Empire and the Transformation of an Indigenous World, 1792–1859* (University of

North Carolina Press, 2010), John Trombold and Peter Donahue's *Reading Portland: The City in Prose* (University of Washington Press, 2017), and Brad Schmidt's 2013 *Oregonlive* article "East Portland's Housing Explosion Tied to City Plan Without Basic Services."

ACKNOWLEDGMENTS

I am incredibly grateful to many people who helped with the publication of this book, and to others who encouraged and supported me during the long process of writing it.

Thank you to excellent editors Nicola Mason and Sarah Haak and to designer Barbara Bourgoyne, all at Acre Books.

Thank you to champion publicist Cassie Mannes Murray.

Thank you to Jill Christman for editing and publishing my essay called "The Polaroid Baby and the Shape of Time" in *River Teeth*, and thank you to Liesel Hamilton for editing and publishing "The Ghost Road" in *The Southeast Review*. These first serial publications helped me persist.

Thank you to Emma Copley Eisenberg for a crucial reading.

Thank you to friends, co-conspirators, and supporters Olivia Cronk, Nicholas Hengen Fox, Sonia Greenfield, Jenine Fisher Harris, Blake Hausman, Joon Ae Haworth-Kaufka, Meredith Kimi Lewis, Jaye Nasir, Christopher Rose, Megan Snyder-Camp, Emily Strelow, Philip Sorenson, Sara Wainscott, and Nadia Wallace.

Thank you to Rochelle Mollen.

Thank you to the Portland Community College Federation of Faculty and Academic Professionals (AFT Local 2277) for negotiating a contract that includes the job-share arrangement that enabled me to write this book.

Thank you to my father, George Johnson, for unconditional support and acceptance.

Thank you to my mother, Leslie Johnson, for the incredible gift of your words, life, and permission.

Thank you to Kevin Edwards, my partner in art and life, for persuading me again and again to keep going.